THINK HAPP

First Published in the UK in 2018 by Independent Publishing Network

A CIP catalogue record for this title is available from the British Library.

ISBN: 978-1-78926-150-9

Front cover and design by Book Printing UK.

Printed by
BOOK PRINTING UK

Remus House
Coltsfoot Drive
Woodston
Peterborough
PE2 9BF

Printed on FSC accredited paper
from recycled, renewable and responsible sources.

THINK HAPPY PLANET

With heartfelt thanks to:

The brilliant peeps referenced at the back of this book for their informative and inspiring articles.

Nascita Montuschi for proofreading, editing, her positive feedback and support.

Susan Dearest for her helpful suggestions, support and encouragement.

Dearest Hopper for her adamantine support and encouragement, and her unbridled willingness to have a good laugh with Master.

My good mate Rach for her encouragement and for listening to me blah on.

Glennie and Belinda for their time and valuable feedback.

Freya for her Scanner Skills and Techy support.

And you. Thank you for reading this book.
I hope it leaves you with a smile on your face.

And finally, I'd like to thank life.
Long may it live.

For my darlings
Jes, Natty, Emrys and Raffi

Love is eternal

CONTENTS

CHAPTER ONE

THINK HAPPY

Happiness depends upon our selves

Aristotle

THINK HAPPY

THE BIGGER PICTURE
We are living in probably the most exciting point in human evolution since we discovered how to harness fire.
We're becoming a Global Community of Guardianship.
A global community realising that our inter-dependence and our co-operation underlie our existence and our survival.

We've unwittingly created for ourselves the ultimate challenge in the shape of our changing climate, with extreme weather and extreme adversity. Happily, adversity unites us, regardless of creed, colour or distance because above all else, we are a social species. The extreme weather patterns and events are bringing us closer together, as one human race on one shared planet.
Climate change is making us think very differently.
It's making us think for the better, making us think for the long-term, making us create the ways and means for sustainable, worldwide wellbeing.
A vital part of worldwide wellbeing is happiness, our personal and our collective happiness. Happiness is a tree of life that bears the vital fruits of inspiration, innovation, connection, motivation and cooperation.

THE SMALLER PICTURE
If there's a time when the world needs more happiness, more inspiration, more connection and more cooperation, it's now.
In our personal lives many of us experience too much stress and too much pressure from the demands and pace of 21st Century living.
Plus there's the collective stress and tear of climate change, war, tragedy, death, of displacement, homelessness,

poverty, hunger, and disease. There's an awful lot of stress about.

All this stress and pressure is taking a big toll on our thinking, our hearts, and our health, and sometimes happiness can seem as elusive as a butterfly.

Injecting more happiness into the world under these conditions might seem a tall order, but the simple truth is that it does start with you and me. By training our brains and cultivating a fresh way of thinking we can bring authentic, lasting happiness into our lives each and every day, we just have to be willing to think a bit differently.

RESPONSE

When we're feeling happy life is sweet. People, situations and inanimate objects tend to fall easily into place and when they don't, we are inclined to take things in our stride. We look for possible alternatives, positive outcomes and won't be upset by the circumstance, event, person or thing that isn't falling into place. Happiness brings us resilience to cope with difficult situations and we don't get so easily stressed, we find a resolution or find a way through. When we're feeling happy we are inclined to look on the bright side of life. It's not the same story when we're feeling tired or unhappy. Life, people, events tend to irritate and annoy us, get in our way and interfere with our schedules and expectations. If we're in a bad mood or feeling tired when something goes wrong, we are easily stressed with the situation and can take it too personally. We can make assumptions about all manner of stuff while we jump over reason to conclusions of yet more stress and pessimism. Sometimes, it can seem as if all the 'things' have joined together to confirm that life's a bitch and then we die.

'Things', however, are not the cause or reason for our happiness or unhappiness. For example, if a car breaks down when we're travelling somewhere, or a train or coach is delayed, the events that unfold afterwards can go either way depending on how we respond. If we're in a good mood we don't immediately think the worst, rather we deal with the situation as best we can and possibly use it as a chance to have a drink or grab a bite to eat. If on the other hand, we're in a bad mood or feeling tired when the car breaks down or the journey delayed, our initial response is probably to give voice to our grumpiness, proclaiming the worst case diagnosis before launching into a long moan listing all the negative consequences and how it's going to impact on the schedule for the rest of the day and week ahead.

The same situation can elicit a different response and outcome depending on how we think and how we feel. We've probably all reacted in both ways to a delay in our journey, either with good cheer or with grumpiness and stress.

It's the same with relating to other people. Often we think and project the source or reason for our happiness as coming from other people. While we certainly do influence each other with happiness, the source of our happiness is from within our selves. We can think and project the reason for our upset and unhappiness onto the very same people that we perceive as being the source of our happiness. When we start a new relationship, the 'in love' factor overrides all manner of stuff. A behaviour in our loved one might seem endearing at first, but a wee way down the line that same behaviour in our loved one can drive us absolutely nuts.

POINTY FINGER SYNDROME

In our society we are quick to blame others for how we are feeling and collectively we seem to have developed a 'Pointy Finger Syndrome', that twitches about looking for where to place the blame whenever things go awry or upset our feelings. Thinking in this way creates a mist of illusion and prevents us from being able to see the true source of happiness. If we think that other people are to blame for our unhappiness, we can get stuck on a foggy road with more unhappiness in our lives. We can't change other people.

The root of happiness doesn't come from what other people say or do, the roots are entirely self-manufactured by how we think and how we respond.

Aristotle said a wee while ago, "Happiness depends upon ourselves" and these words are just as helpful today, two and a half thousand years down the line. Happiness does depend upon how we think, feel and respond to any situation in life, including a delayed journey.

I reckon Aristotle would have been very impressed with the talented musicians who, when stuck in a standstill traffic jam on the M5 (Sat 12th Sept 15), decided to get out their car and play 'Pacabel's Cannon' to their fellow stationary, and delighted, travellers.

NO BACK-SEAT DRIVERS

Thoughts lead to feelings, which lead to action and response.

Often, because life is too busy and too full of stress, we react negatively to stuff without even thinking very much, it's as if we don't have a choice, we just react. Even though we do have a choice of which response-road to travel on, frequently we ignore the signposts or we just don't see them, our buttons have been pushed and we're on a slip road heading for Stress City.

In order to stay on the right road we need to learn to drive the car a bit differently, use our 'brain brakes' more often and with both hands on the steering wheel, acknowledge that there are no back seat drivers. It isn't anyone else's fault if we're driving on the wrong road. No matter what other people think, say and do, how we respond is entirely up to us. Acknowledging that we are fully in charge of our individual happiness gets our brains thinking in a different way.

Taking full responsibility for our thoughts and feelings is the first step in gaining our advanced life-driving licence: the ability to respond differently.

When we're heading off down the wrong road and the Pointy Finger starts to twitch, we can train our brains to activate the warning lights on the dashboard, change gear to neutral and stop the car. We need to clear the mist of illusion from the windscreen, tune the radio to Think Happy and select a different route of thought.

21ST CENTURY PRESSURE

One of the major components getting in the way of our individual and collective happiness is the fact that life today is so very busy.

We're under a lot of pressure in the 21st Century from all angles of life to buy into and believe a false equation for happiness. We're under pressure to work long hours, pressure to look a certain way, travel abroad, drive a certain car, keep up with all the blah on telly and online, and to have the very latest technological devices, all of which society promotes as necessary for a successful and therefore happy life. In real terms too much pressure leads to more stress, poor health and definitely less happy.

We're easily caught in the fast lane of pressure and stress and can travel for miles and miles before we realise that we're going in completely the wrong direction for happiness.

Our society promotes the ultimate in success and happiness as having a job that pays extremely well. Most of us probably don't have jobs that pay extremely well, are under pressure to work long hours, and happiness gets scheduled for a couple of hours at the weekend.

If we take on and believe society's equation for happiness that Happiness equals Long Working Hours plus Money: $H = LWH + M$, most of us end up getting the maths wrong. For quite a lot of us the equation looks more like this: $LWH = PS + PH - H$, Long Working Hours equals Pressure and Stress plus Poor Health, minus Happiness.

We are constantly bombarded by the advertising industry with messages to buy more and yet more stuff. We're persuaded that having an array of different products will ensure a successful and happy life.

We're sold loans to pay for it all and then we're sold insurance to protect all the stuff we've bought. We keep spending because society and the advertising industries have led us to believe that we can 'Buy Happiness'.

We're persuaded by adverts, we're enticed by the shops and we're backed by society to spend money even if we haven't got any. We've been engrained to believe the equation $H = LWH + M$, and because we want to be happy, we want to be and be seen as successful, we spend the money and we buy the stuff.

FLEECED

Advertising is a highly successful industry that's very good at its job.

Shops are also extremely good at getting us to buy items we don't need, and more food than we can eat. We're all familiar with the way supermarkets will jiggle the lay out of the shop so that we have to search for the items we want to buy. We're forced to look at other products, which slows us down while we try to find the things we do want. Jiggling things around in this way works really well for the supermarkets because invariably we end up buying items we don't need and that weren't on the list.

I heard on the radio that a big UK department store (can I say John Lewis?) purposefully slows the customers down on entry into the shop. The foyer is very welcoming and set out in a way that immediately creates a sense of relaxation, warmth and comfort. It's peaceful away from the noise and bustle of the street and we're made to feel at home. We can see across the shop and we can see the escalator, but the shop floor is laid out in such a way that it's hard for us to get anywhere directly; the layout is physically designed to slow us right down. If we've slowed down, we'll look at more products, we'll browse around and in the absence of clocks, we'll take our time and we'll buy more stuff.

If one department store is using floor layout and design strategies to slow us down and coerce us into buying more, then most shops are probably doing the same thing: driving us like sheep into a pen to get fleeced.

It's easy to spend money nowadays and easy to spend money we haven't got with the help of plastic credit cards. Our society promotes and perpetuates the belief that it's normal way to buy items even if we haven't got the money, and consequently, the belief that it's normal to have a lot of debts.

More debt means more stress and less happy.
The 'have what you want' way of thinking promoted and perpetuated by society and advertising corporations, together with the false notion that we can buy happiness, has engendered a consumer attitude of greed amongst us; an attitude that is driving us, and the planet, into crisis and a very long way from happiness.

SHELF LIFE
It's human nature to want comfortable lives but do the items we're persuaded to believe will bring happiness actually make us happy? How long does the feeling of happiness last after we've bought something? A day? A week? A month?
What happens once the happiness effect has worn off? Because we've been taught to 'Think Consume' we're easily persuaded to go and buy more things. The associated happiness factor fades because the source of happiness doesn't come from the things we buy despite how many adverts proclaim otherwise.
We can't buy happiness.
Consumerism is one big, carroty illusion of happiness and we've been tricked like blinkered donkeys never quite feeling satisfied.

SLEEP
There has been enormous change during the last 150 years since the Industrial Revolution which, together with the technological explosion has radically altered the way us humans go about living human life.
Many companies work round the clock, keeping their product or service continually available to buy, many shops stay open late or all night, requiring many people to work silly hours and night shifts.

With electricity our evenings have become much longer and come to be regarded as 'Extra Daylight Hours' during which we consume more and stay up late. It's become normal to live our lives in this way.

The amount of pressure we are under in our busy days and long nights lifestyle is having a detrimental impact on the quality of our sleep.

Sleeping well is fundamental cornerstone for good health and an imperative ingredient for happiness.

The amount and quality of sleep we have has a direct impact on our minds, bodies and wellbeing. After a good nights' sleep we wake feeling refreshed, positive and happy whereas if we haven't slept well, we're likely to wake up feeling physically tired and groggy, maybe feeling emotionally drained and definitely not the happiest person in the shower.

We carry on anyway even if we haven't slept properly for days or weeks or months because the bills have got to be paid and there's all the stuff to buy. So we drink tea, coffee, energy drinks and whatever else to keep us going. Later we drink alcohol and whatever else to help us relax and switch off from the pressure. Then we take sleeping pills to help us switch off the stress in our minds in order to get some sleep.

Insufficient sleep has a significantly adverse affect on our physical and mental health. According to Professor Mathew Walker, neurologist, sleep scientist and author of 'Why we Sleep' (published by Penguin) says that sleep is more important for our health than diet or exercise, and very few of us are getting sufficient sleep. Insufficient sleep not only drastically impairs our health it also affects GDP. When we're tired we choose easier options, we're less productive and we're less creative, whereas good sleep produces a threefold Increase in creativity. Our commuting times are longer; we're leaving home earlier, getting back later and stay up later. Working patterns have changed to reflect the rise in consumerism and the expectation for instant delivery within 24 hours, which relies upon many people working night shifts. The World Health Organisation has recently defined Night Time Work as carcinogenic.

OVERLOAD

Not responding to the real needs of our brains and bodies wrecks the immune system and lowers our resilience to the demands of 21st century living. We become more stressed and our health, and our happiness suffer. Stress creates a vicious circle of stress.

Often, if we do start to have problems with our health we carry on regardless because of the pressure, work, bills and all the stuff to buy and in the long run this can take a big toll on our health and we'll be forced to stop, get help, get rest and get better.

I heard on the radio that more people in the UK, when compared to the rest of Europe, die from diseases that could have been treated if only they had sought medical help sooner.

The NHS is a fantastic service currently under huge pressure due to a massive workload and under-funding. There isn't enough money or staff to cope with the vast numbers of the UK public who need medical help. With so much pressure and stress in our lives we've hit overload. So many people needing medical help is a big indicator that our society has got the wrong equation for happiness.

On the bright side, we've got the gift of flexi-wiring and we're designed to keep re-thinking.

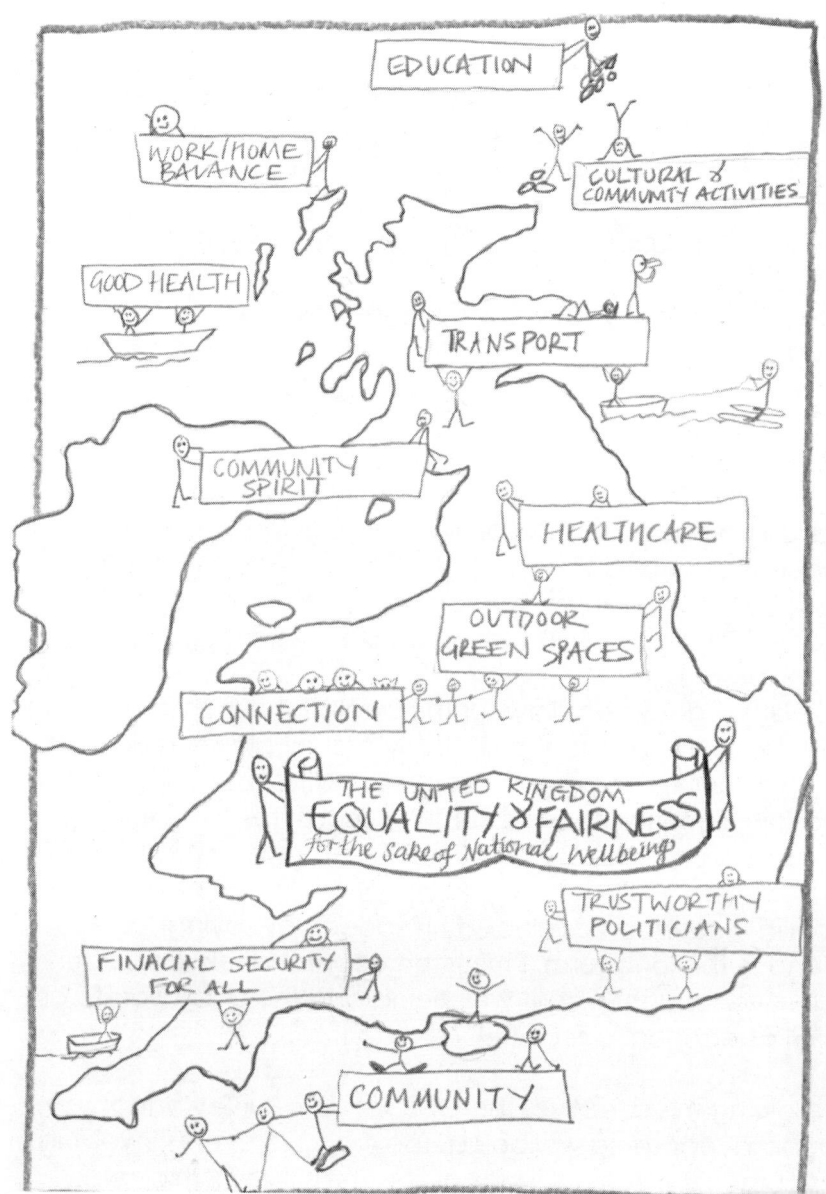

UK SURVEY
In November 2010, under the coalition government, David Cameron made a speech on Wellbeing to announce that the cabinet were asking the Office of National Statistics (ONS) to 'devise a new way of measuring wellbeing in Britain'.
Here are a couple of excerpts from the speech:

"From April next year we'll start measuring our progress as a country, not just by how our economy is growing, but by how our lives are improving; not just our standard of living but by our quality of life."
He also said "We'll continue to measure GDP as we have always done. But I do think it's high time we admitted that, taken on its own, GDP is an incomplete way of measuring a country's progress."
David Cameron then went on to quote from a speech that Robert Kennedy gave forty years ago regarding GDP:

"(GDP)... does not allow for the health of our children, the quality of their education, the joy of their play. It measures neither our wisdom or our learning; neither our compassion nor our devotion to our country; it measures everything, in short, except that which makes life worthwhile."

David Cameron concluded his speech by saying:
"And in the long term, I think the country would be better off if we thought about well being and the quality of life as well as economic growth." (1)

Following this directive, the ONS set up a survey and debate regarding wellbeing. They ran 175 events involving 7,250 people from all walks of life. The survey, which was called 'What matters to you?' generated 34,000 responses, some of which were from organisations and groups of people representing the opinions and views of thousands more.

The survey demonstrated that people in the UK have much in common. The things that matter most are: the importance of good health for our wellbeing and happiness, the need to connect with people and have good relationships at home, at work and in our local and wider communities, and the importance of the environment around us.

All age groups emphasised the importance of these factors together with financial security, equality and fairness as being a vital part in wellbeing for us as individuals and vitally important for the country as a whole.

Other important factors highlighted in determining our personal wellbeing are: having a balance between work and home, time outdoors and availability of green spaces, participation in cultural or creative activities, availability of government services such as health care, education and transport networks. Also highlighted is 'a greater sense of fairness and equality for the sake of national wellbeing'.

Many people mentioned that 'community spirit' is a crucial factor for every ones' wellbeing but it's currently lacking in our communities. Many people also highlighted the 'importance of democracy and having accountable, trustworthy politicians'.

The results of the ONS survey shows that wellbeing is of the utmost importance in life. People in the UK want to be happy, want to belong to communities, want fairness for all and we want a method to measure wellbeing in our country. (2)

WORLD SURVEY
It's not just us lol in the UK that value wellbeing and want to be happier. In 2006 Nic Marks, the founder of the 'Centre

for Wellbeing' based at the New Economics Foundation (nef), published the first 'Happy Planet Index'. Their website describes how the message 'resonated with hundreds of thousands of people around the world and within two days of it's launch the report was downloaded and read in 185 countries worldwide.'
That's a lot of countries interested in having a happier planet. With currently 195 countries worldwide, it equals 94.87%.

In July 2010, Nic Marks gave a talk with the feedback, statistics and information from the first global survey 'Asking The World What We All Want?' (The video is on the nef website along with fantastic articles by inspiring people working for a better, happier future for all.)
The feedback clearly demonstrates that above all else, and across the whole world, billions of people value happiness as the most important quality in life.
The graph in the video shows happiness at the top, followed by love, then health, and a sizeable chunk below comes wealth; clearly illustrating that money isn't the main aspiration for most people around the world.
Nic Marks said: "People all round the world say that what they want is happiness, for themselves, families, children, communities. Money is important, it's there, but it's not nearly as important as happiness, it's not nearly as important as love, we all need to love and be loved in life. Money's not as important as health, we want to be healthy and live a full life."

Happiness comes out tops for what the whole world wants most.

THE HAPPY PLANET INDEX

The Happy Planet Index is a way of measuring a nations progress in wellbeing rather than how much stuff and money we have. It measures how successful a nation is at 'creating happy and healthy lives for its citizens'.

The main criterion in calculating an index for each country is how much of the planet's resources each country uses. As Nic Marks says, in economic terms the planet is the "ultimate scarce resource – we have only got one planet and we all have to share it".
Economics examines how to get the best desirable outcome for the resource. The best desirable outcome for our ultimate resource, the Earth, is for people to live happy lives while using as little of the planet's resources as possible. After all, we all want there to be some planet left for our children's children and generations to come.

The HPI calculations are illustrated on a graph with Eco-Footprint plotted horizontally and Happy Life Years, quantity and quality, plotted vertically. The graph clearly illustrates how happy we are across the whole world together with how much of the planets resources we are using in order to be happy.

The 2006 report showed that the Latin American countries are extremely good at being happy without using loads of the planets' resources. Costa Rica came first as the happiest country in the world and using only one quarter of the resources used by a typical country in the western world.

There are a few differences about life in Costa Rica that give it such a happy status. For a start, it's got the Caribbean Sea on one side, the Pacific Ocean on the other and it's near the equator so I reckon they have a bit more sunshine than we do in the UK. Along with better weather, Costa Rica generates 99% of its electricity by using renewable resources and their government was one of the first worldwide to commit to being carbon-neutral by 2021. The average life expectancy in Costa Rica is 78.5 years, which is higher than the average life expectancy of a citizen of the USA.

Another significant fact about Costa Rica is that in 1949 the government abolished their army and invested the money instead in improving health and education. They now have the highest literacy rates in Latin America and in the world. Nic Marks pointed out that "Costa Ricans have the Latin vibe! They have social connectedness" and he went on to say: "the future might not be North American, it might not be Western European, it might be Latin American." (3) Which would suit me right down to the ground; more sunshine, no wars, no armaments, clean power, happy connected people and yes, men who can dance.

GLOBAL CHALLENGE
As well as learning to dance (which is one of the activities scientifically proven to be excellent for keeping us mentally fit and happy in our older years), the collective challenge for the world is to raise the Global Average on the HPI so that we all have higher standards of wellbeing with a lower use of the planets' resources.

And we can be happy without it costing the Earth.

The UK has made some progress since the first report of the HPI in 2006 where we scored a shameful overall ranking of 108th out of 178 nations.

In 2009 we ranked 74th place out of 143 nations and in 2012 we moved up another notch and ranked 41st out of 151 nations. (4)

Although this progress is encouraging, on the flip side, there's still a long way to go yet. The placement of the UK at 41st place in this report put us behind many countries in the developing world.

We have the 18th largest eco-footprint worldwide and 'If everyone had the same eco footprint as the average citizen of the UK, we would need nearly three planets to sustain us.'

The USA also has room for significant improvement.

They ranked 150th out of 178 nations in 2006, 114th out of 143 nations in 2009 and 105th out of 151 nations in 2012.

These results clearly show how us lot in high-income countries impair the level of world happiness by consuming so much of its resources and having such a high ecological footprint. (5)

In the most recent Happy Planet Index of 2016, Costa Rica are in first place again, continuing to set a shining example to the world on how to be happy without costing the Earth, and without the need for war or armaments.

Their HPI overall score was 44.7 with an ecological footprint of 2.8.

The 2016 Index shows that the UK moved another notch and ranked 34th place out of 140 nations, with a HPI score of 31.

Which on the face of it looks encouraging, however, the UK ecological footprint is still shocking, we're placed 107th out of 140 nations.

Along with our American friends, we really do need to pull our socks up. The USA ranked 108th out of 140 nations, HPI score of 20.7, and an even more shameful ecological footprint placement of 137th out of 140 nations, which won the USA an award for being in the bottom 10 of world ecological footprints.

Which begs the question: Why is the USA constantly referred to as the most powerful country in the world? The 'most shameful country in the world' would perhaps be a more appropriate title.

On the good side, now that we have international measures and debate regarding wellbeing and happiness for us all, one bunch of people sharing one precious planet, we could progress very quickly. The Happy Planet Index has made our consuming habits as visibly shocking as the forests that have disappeared.

We do influence each other and nations influence other nations for change. Many countries are following Costa Rica's fine examples and investing in our collective future with clean, renewable energy, wellbeing and education. It's possible too that instead of investing in the production of war and armaments more countries will follow Costa Rica's model and choose to invest in life and happiness.

THE JOE BLOGS: DAILY PROGRESS REPORT
The OECD, the Organisation for Economic Co-operation & Developments, have recorded data since 1960 for some of the countries in the HPI group. The trends between 1960 and 2015 show that there has been a slight increase in wellbeing but this slight increase has come with a massive cost to the Earth's resources.
Nic Marks said: "overall, we have been very inefficient at looking after either ourselves or the planet".

He went on to say that we need instant feedback in our homes such as smart meters (non-toxic variants) showing how much energy we're using and what it costs, and we need to create a collective goal; to raise awareness and collectively achieve the UK's target of 3% carbon reductions a year. He suggested a daily report stating how much closer we are to our collective target for wellbeing and our collective targets for carbon reduction, just as the Dao Jones world stock exchange reports daily about how much one currency is worth against another. It's a great idea and would be far more relevant to far more people than the Dao Jones money thing. It could also be published on social networking sites where we could engage, connect, be inspired and motivated.

Money isn't top of the list for what the world really wants. Having a highly visible, positive goal for what we do want, something that we do relate to and can check in with every day could take off and fly. 'The Joe Blogs' would change our collective mind set. With daily images illustrating how successful we are as a nation at improving our wellbeing and collectively reducing our carbon output would make a massive difference to our motivation, enthusiasm and willingness to engage more positively, think and do things differently.

The Joe Blogs could substantially influence us in the UK in meeting our targets while injecting us with a collective sense of community, purpose, wellbeing and happiness.

US LOT

Meanwhile, while we're waiting for The Joe Blogs Report and waiting for an update on 'The Wellbeing of the UK' (which won't be forthcoming from Dereliction Dave, his idea of wellbeing seems to be to dump us all in the stormy Brexit Seas while disembarking the sinking ship faster than you can hum a little tune), to begin to make a difference in the world and invest it with more happiness, we can just get on with it ourselves.

How do you rate your current level of happiness?

On the Happy Planet Index website there's a personalised test to assess our own level of wellbeing, which includes a calculation for individual eco-footprints. It's worth checking out and having a go.

Taking 'snapshots' of different areas of our lives gives us a moment to reflect on which aspects are lacking in a sense of happiness or heading down the wrong road.

This exercise can be done by drawing a chart of the different areas, each with a score out of ten. Getting a clear overview is very helpful and useful to refer back to when measuring progress further down the line. Or you could take a few minutes now to reflect on the following areas of your life, a 'snapshot' of the moment.

HOW HAPPY ARE YOU?
How happy are you with the following:

Your general level of happiness?

Your general level of health?

The amount of love you experience with family and friends?

The amount of love you experience with your spouse or partner?

Your job satisfaction?

Your financial situation?

Your local environment?

The larger environment?

Education & training for yourself, family or friends?

The amount of fun you have in your life?

Your participation in the local community?
The amount of creativity in your life?
The time you spend volunteering?

Your own eco-footprint?

It's a useful list to highlight the areas of our lives where we could think differently, act and respond differently in order to create more room for happiness.

SOURCE
Situations, events and other people are not the cause of our happiness or our unhappiness; it's how we think that makes the difference. With full acknowledgement that we are in the driving seat of our own happiness and with both hands firmly on the steering wheel, we can train our brains to Think Happy.
We can learn how to retreat from the fast lane of pressure, stress and carrots. When we're heading off down the wrong road and the Pointy Finger starts to twitch, we can train our brains to activate the warning lights on the dashboard, change gear into neutral and stop the car. We can train our brains to select different, helpful routes of thought.

We can find encouragement from the collective spirit of the UK wanting greater wellbeing and fairness for all.
We can take solace from and be emboldened by the fact that the whole world values happiness as the most important quality in life.

Fuelled with enthusiasm and motivation by this knowledge we can inspire and influence our world with positive thought, action and a heap more happiness.

Think Happy.

CHAPTER TWO

THINK THINK

As we sow, so shall we reap.

Proverb from the Bible

THINK THINK

What we think about ourselves, and the way we talk to ourselves internally, are the most crucial factors in determining the amount of happiness we experience. By training our minds to think differently we can unlock our potential for greater, lasting happiness.

PERSONALISED GUIDE BOOK TO LIFE

What we think creates who we are. We are the sum total of all our thoughts: all the good thoughts, the upsetting thoughts and all the thoughts in between.

We are what we believe of ourselves, positive beliefs about what we do well, what we like, what works in our life and what we value, together with what doesn't work, what we don't like and any limiting beliefs, all these thoughts together make for who we are.

This, together with what we believe and value in our external lives with family, friends, colleagues, communities, the wider society and the world at large, all combine to create our own individual interpretation and understanding of reality. All these thoughts, values and beliefs combine together creating a 'Personalised Guide Book to Life' through which we understand ourselves and engage with other people and the world around us. We live in the same world but we all have an individual Guide Book because we filter information, events, people and absolutely everything according to our own perception, beliefs and values.

We have a lifelong collection of beliefs and values, much of which we collected from our immediate family, friends, school, and locality as we were growing up. The beliefs and values about ourselves that we have collected by the age of seven become embedded into our unconscious minds.

age we don't really choose our beliefs, we just soak
ıat we are told.
Iⅼᵤₑe first few years of our lives are when our brains are
constantly loading all the software onto the hard-drive and
it's possible that we hang on to limiting beliefs about
ourselves that can hold us back later in life, or hang on to
beliefs that weren't really ours in the first place. We can
weed out these beliefs and values that no longer help or
that limit us in some way; we just have to find and identify
them first.

Some of these beliefs and values we hang on to because
they're wrapped up in our sense of identity, and because
we are influenced by what other people believe and value.
Some of the content in our Personalised Guide Book to Life
isn't necessarily what we actually believe and value about
ourselves, or the world around us.
To begin to think differently, we need to look at how much
of what we think, believe and value genuinely makes us
feel good and happy with ourselves, and happy with the
people and the world around us.

ELECTROMAGNETIC ONESY
All the thoughts, beliefs and values which create our
individual Guide Book to Life, exist internally in our minds,
exist in our bodies and exist externally in the electrical and
electromagnetic fields around us. Our Guide Books
surround us in an energetic form like an invisible,
electromagnetic, loose fitting, and extremely comfortable
Onesy, unique in colour, style and design, which is
constantly sending and receiving information.

The energy fields of the invisible Onesy enable us to sense
how other people are feeling and sometimes we can
actually feel what another person is feeling. Other times

we're with someone who is pretending to be okay and saying the right words, but we know they're not okay because we can sense it.

We've been led to believe that we only have five senses: sight, touch, sound, taste and smell. What about the sense when we just 'sense' something? This is generally referred to as ESP, but the problem with calling it Extrasensory Perception is the suggestion that it's 'an extra' and only available a select few, which is just plain silly. We all have sensory perception because that's exactly what our electric and electromagnetic fields do; our invisible Onesies work like radar equipment and are constantly perceiving and interpreting the energetic information around us. It's not 'an extra' or 'an app' that only some can download. It's as much part of each one of us as our eyes, ears, mouth, nose and skin, and these Onesies are an extremely important part of us even though we can't see them.

Everything we think, feel, believe, value and experience is woven into the fabric of our Onesies. Our thoughts, feelings and experiences are the threads of the invisible, electromagnetic fabric that surrounds each of us.
Science today confirms everything is connected. These electric and electromagnetic Onesy fields connect us to each other, to our environment and the Earth. The Earth is connected to our universe and our universe is connected to the big beyond.
I'm not so fussed with the big beyond; this universe is big enough for me.

LAW OF ATTRACTION
There's a well-known proverb that's based on a passage in the Bible: 'As we sow, so shall we reap'. Whatever thoughts, feelings or actions we invest in, whatever seeds we plant,

will bear fruit accordingly. This notion is described in current lingo as 'The Law of Attraction'; whatever we give out we attract back to us. In the Law of Attraction, like attracts like. Whatever it may be, the good, the bad, the beautiful or the ugly, whatever you're putting out; it's coming right back at ya.

Whatever we think, feel, value and experience creates who we are.
It creates our Personalised Guide Book to Life, every minute detail of which is held as energetic information in the fabric of our Onesies. Whatever we are wearing in the electromagnetic fields of our Onesies is exactly what we attract to us. We attract into our lives more of what we already think, feel, believe and value. There's no discrimination. We attract responses to all the positive thoughts, beliefs, actions, delights and surprises that are woven into our Onesies just as much as we attract responses to all the negative thoughts, limiting beliefs, worries and fear.

The Law of Attraction is as powerful on this Earth as the law of gravity. What goes up must come down, what we radiate out comes back.
In order to change what we attract into our lives we have to begin by changing our thoughts.
Brains like having to re-think things, it's one of their favourite pastimes so we can easily learn to think a bit differently.
We can review and rewrite our individual Guide Books and mend the holes in our Onesies.
We can take an objective look in our Cupboard of Thoughts, the Wardrobe of Beliefs, the Cloakroom of Values, and have a thorough spring clean, discarding all the stuff that doesn't fit or doesn't suit us anymore. We can ditch the stuff we've inherited and which wasn't ours in the first place

and slowly but surely discard the thoughts, beliefs and values that no longer serve us or bring happiness.
We can wash our Onesies at 90 degrees and hang them out to dry in the fresh air. (If you're feeling a tad under par however, I recommend the delicates wash and a slow spin).

CHEMICAL PACKAGES
Every thought and emotion we have generates a response in our brains, which triggers corresponding electrical and chemical messages throughout our bodies. When we're feeling good the brain sends positive electrical and chemical messages associated with the good feeling throughout our bodies. Likewise, negative thoughts, emotions and words generate electrical and chemicals messages filling our bodies with the associated chemicals of worry, stress, doubt or fear.

"I AM" THE SUBCONSCIOUS MIND
It's hard to over-ride the subconscious mind without first deleting the relevant limiting thought programmes that are already installed and running in our minds. The subconscious mind might appear to be a back seat driver but it's this part of our mind that really is the driving force and will do everything in its power to back up the words we use, the thoughts, beliefs and values we hold, whatever they may be.
A helpful road into the labyrinthine workings of our minds is to begin to notice the small and subtle inferences of the words that we use.
Sentences that begin with "I am..." are very powerful in determining who and what we are. We start a lot of sentences with "I am or I'm..." and we need to pay careful attention to the words that follow.
A simple example, if we repeated say "I'm tired of..." a certain situation, that's exactly what we become. We

become physically tired and have less resilience to deal with it. If we think or say "I'm tired" because we really are feeling tired, by repeating it either out loud or internally will compound things even more. The subconscious mind will respond to whatever we say we are and will transmit the chemical messages associated with tiredness throughout our brains and bodies. By repeatedly thinking or saying "I'm tired" we become more and more tired until eventually we fall asleep at the office desk with the keyboard imprinting strange hieroglyphics on our forehead.

Whatever we say and think, positive or negative, our subconscious mind will set to work backing it up on the hard-drive; it's part of its job description and our subconscious mind is exceptionally good at its job.

To think differently and sustain positive thoughts and feelings, we have to be a bit picky and carefully select what words to use in our internal dialogue. We can start to compile an inner drop-down Options Menu with alternative thoughts and vocabulary. We can actively tune the radio in our minds to sing a different song and choose thoughts and words that inject our bloodstreams with life-affirming chemicals. We can gradually improve our driving skills in how we respond to the pressure and demands on the highway of life, and to its rewards.

NEURAL PATHWAYS

We are constantly responding to the environment around us. Every response we have travels down an established pathway in our brains or creates a new pathway, stores information to memory and retrieves information from memory.

The sensory information we receive from all of our senses, sight, sound, touch, smell, taste and Onesy, is sent to our brains via electrical and magnetic pulses in our neurons, which together create neural pathways in our brains. These neural pathways become more defined the more they are used; the more we think the same thing or do the same action. If we repeatedly respond to a certain stimulus in the same way it becomes easier and easier for the neurons to 'fire' along the same route and for our response to the stimulus or situation to become much quicker. Learning to ride a bike, drive a car or play a musical instrument are good examples of how we learn. The first time we have a go there's a lot to think about, a lot to remember and we're acutely aware that we're learning a new skill. The more we repeat the new skill, the easier it becomes and eventually we aren't even consciously thinking of those things which at first seemed so very difficult. The learning information gets ingrained in our brains, stored to deep memory in our subconscious minds and our responses become automatic. The formation of these new pathways in our brains is a bit like the way water will carve itself a path through the land; the more water that continues to flow the deeper the channel is carved. Developing our brain skills to think differently works in exactly the same way.

TAGGED

Every response we have is accompanied with an emotion that gets 'tagged' when the information is stored to memory.

Sometimes responses to certain stimuli become instant reactions, there's hardly any thinking involved, we just react. These 'tagged' reactions can be either positive or negative. When we have an instant negative reaction often the Pointy Finger starts to twitch looking for somewhere to place the blame for our feelings.

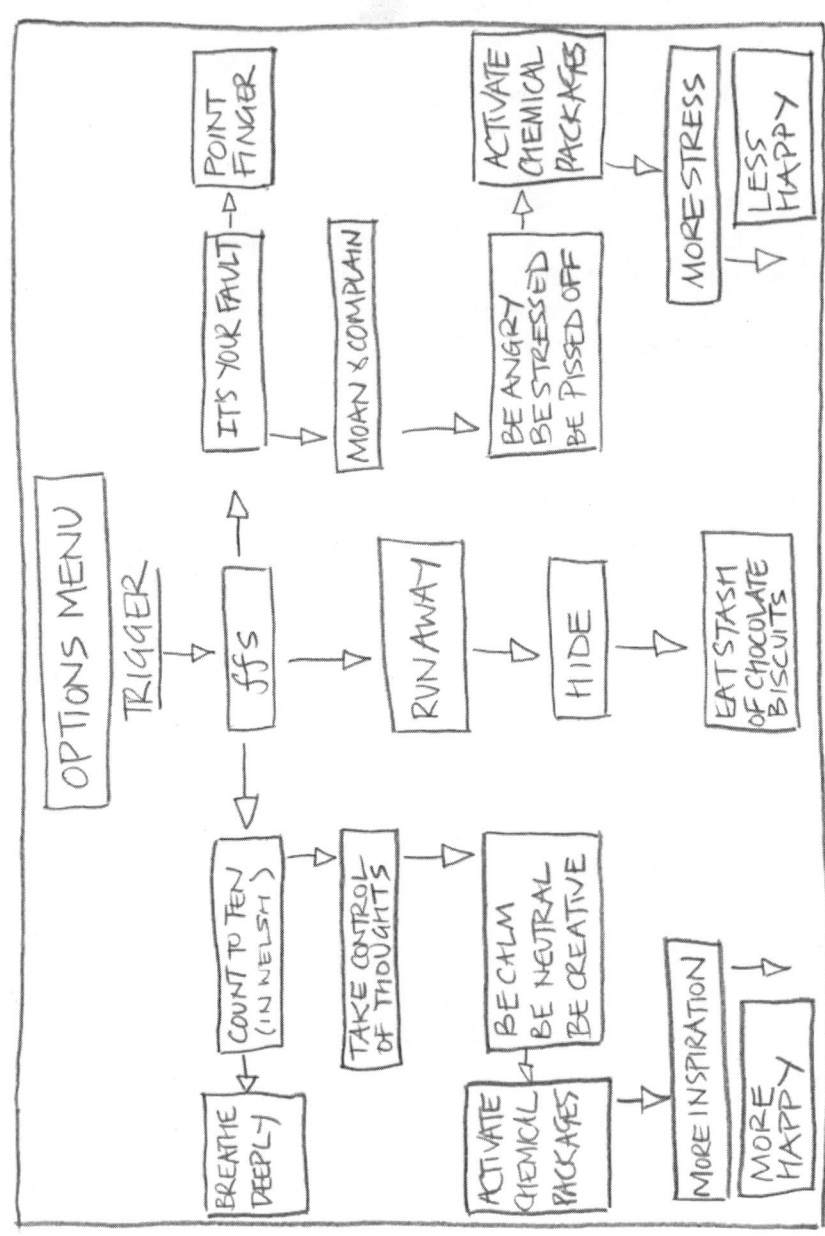

OPTIONS MENU

TRIGGER

ffs

IT'S YOUR FAULT → POINT FINGER

MOAN & COMPLAIN

BE ANGRY
BE STRESSED
BE PISSED OFF

ACTIVATE CHEMICAL PACKAGES

MORE STRESS

LESS HAPPY

RUN AWAY

HIDE

EAT STASH OF CHOCOLATE BISCUITS

BREATHE DEEPLY

COUNT TO TEN (IN WELSH)

TAKE CONTROL OF THOUGHTS

BE CALM
BE NEUTRAL
BE CREATIVE

ACTIVATE CHEMICAL PACKAGES

MORE INSPIRATION

MORE HAPPY

This is when we could pause for thought, Think Think, and try out a different thought from the Options Menu.

By acknowledging that our negative reactions are filling our bodies with the chemicals of stress, upset and anxiety, we've got a starting place and can begin to do something about it. It's an opportunity to think differently, learn how to respond in a way that doesn't stress us out but keeps us driving in the direction of happiness.
The bottom line is: we can't change anyone else.
We can, however, affect and influence one another by changing our own thoughts, our own words and our own actions. We can think a bit differently and learn how to re-set the automatic buttons back to a place of neutrality. By changing our thoughts, words and actions, we change what's encoded in our Guide Books and change what's held energetically in the fabric of our Onesies.
When the Onesy energy has changed, we've changed what we're giving out and consequently what we're attracting right back at us.

FLEXI-WIRING
The neural pathways in our brains are described as having a 'plasticity' quality that makes these pathways very flexible and adaptable. This is extremely handy when it comes to re-thinking anything, especially negative reactions and limiting beliefs.
With our flexi-wiring we can re-think thoughts and beliefs that have become ingrained and stored to deep memory.
A stream of water will change its course to adapt to the landscape it's flowing over; if someone puts a rock in the path of a stream, the water will just flow around it in a new direction.
It's the same principal with the neural pathways in our brains; if we actively choose to think differently about

ourselves, or other people, or situations, new pathways form in the brain and old pathways that are no longer used gradually wither and fade away.

Like learning any new skill, we can learn to think differently about ourselves in the driving seat of our own happiness. With repetition the new thought pathways become more and more established. We just have to practise changing gears, keep our Options Menu fully active and updated, and learn how to put the brakes on if we're heading down the wrong road.

NOISY PLANES

We can't change other people and other people aren't responsible for how we feel. We do have a huge effect upon each other but none the less, whatever or whoever might be the trigger, we alone are responsible for how we feel and how we respond to all that life presents to us.

A few years ago I was living with my family in a beautiful valley in Wales, next to the historic waters of the river Dyfi. We were surrounded by stunning scenery, engulfed in tranquillity and blessed with the immediacy of nature. There were even otters in the river at the bottom of the garden. It was idyllic and felt like my idea of paradise on Earth. What I didn't know when we moved there was that the house and garden were situated below the flight path of The Royal Air Force training route. The planes came as a big surprise and a big nuisance. Sometimes they would fly extremely low, some were immensely powerful and ridiculously fast and filled the valley with a thunderous noise. The noise did not equate with my idea of paradise. Every time a plane flew over I got more and more angry. I started asking in the nearby village about what could be done as it seemed to me that they must be contravening the law and flying too damn low. I was met with resignation. People in the village had been complaining to the authorities for years but the planes just carried on coming in just the same way.

I was in a quandary. A rather pissed-off quandary. I knew it was up to me, I had to respond differently, the planes were going to keep coming and the horrific noise wasn't going to go away. If the sky was cloudy the noise was even worse, reverberating around for ages underneath the cloudbank. This was Wales remember, so the sky was often cloudy and the noise was absolutely deafening. I was stuck and I knew it; I could see the quicksand beneath my feet. The planes kept coming and I kept getting angry over what I believed to be a violation of the sanctuary of my home.

One thing I was acutely aware of during all this was gratitude for the fact that the pilots were training and not actually dropping bombs on my home and village, but my heart went out to the folk of war-torn countries. Although I held on tight to compassion the noise (and purpose) of the planes still really angered me. I tried to think of the planes like Brecht, a playwright who described soldiers as 'our brothers in uniform', but it didn't work; the planes still annoyed me intensely, my brothers in uniform pissed me off immensely and I was still in the quicksand of my own tyranny.

Funnily enough, it all turned around in a moment. It was a beautiful summer's day and I'd gone for a swim in the river. The house was on its own, a good couple of fields away from any neighbours and the river hidden from view, hence I was skinny-dipping. The river was very cold so I was just inching my way in slowly and gradually, the water up to my waist, when I could hear a plane coming. The water was far too cold to dive in, so I stood my ground figuring that the plane probably wasn't flying close by anyway.

But I was wrong. Within about a minute the pilot had flown round the block of mountain and was heading straight back to me. This time ridiculously low, the lowest I had ever seen them fly and right above me. Not only that, he'd radioed one of his mates as a second plane flew in on his tail to get a look at the (now submerged) skinny-dipping woman. My reaction? Did my buttons fire off on autopilot? Did I swear and curse at the deafening noise? Did I get angry and stressed at the proximity and violation of my space? Nope. Not a bit of it. Something snapped, over-rode my hard-drive and seemingly without my consent, I was laughing, loud and out loud. In that moment I finally got out of the quicksand; my response is mine alone. The pilots were just some regular blokes, some regular blokes with a sense of humour.

From that day, I experienced the planes and the pilots differently. The planes kept on coming, the noise was still atrocious, my heart still ached for those in countries at war, but my response had changed. Because of laughing so much, I realised a kinship with my brothers in uniform; we are all just people, whatever we believe. Instead of getting upset and angry I smiled and waved. Instead of filling my body with the chemicals and hormones associated with upset and stress, I filled myself with chemicals and hormones associated with happiness and goodwill.
The planes proved to be a very useful lesson in the end, and I often remind myself of this episode when my Pointy Finger starts to twitch.

INNER VOICE
What we say and how we talk to ourselves internally has a direct impact on our immediate level of happiness. To upgrade our happiness level and begin to think differently, first we have to take note of what we're already thinking and saying in relation to ourselves. Examine the words we use and the tone of our internal voice before we can change them to something else, something better.

The inner voice is constantly talking and this internal dialogue is hard to switch off, but while we learn to calm this part of our wittering mind we can at least change some of the content and delivery.

By changing the content and delivery of our internal ramblings we can begin to harness the state of happiness more consistently. Once we've identified negative or limiting thought patterns about ourselves, thoughts that hold us back, keep us stuck, or thoughts that present hurdles to our own happiness, we can delete them and begin to re-write our own programming. However, as with computer software, if we don't ditch the old version first, the new version doesn't ever quite run properly or it just won't work at all.

THINK OUTSIDE THE HEAD

In the first instance we need to step back a little from ourselves. By stepping back a wee bit we're sort of thinking 'outside the head'. From here it's possible to just listen impartially and become aware of the nitty-gritty of what we say and how we are saying it.

By thinking outside the head and observing from a place of neutrality we gain a clearer understanding of the true nature of our inner vocal soundscape, and the words and phrases we often repeat.

THE NITTY-GRITTY

Observe how much of what you say to yourself is either positive or negative.

Notice how much of your internal dialogue is self-supportive, using words that are encouraging, kind and caring. Notice how much positive feedback you give yourself. Observe how much of your internal dialogue is self-empowering and engenders good self-esteem.

Notice how much of your internal dialogue is critical of yourself, using words that are discouraging, harsh or uncaring.
Observe how much negative feedback you give yourself. Notice how much of your internal dialogue is disempowering and undermines your self-esteem.

TONE OF VOICE
While observing our thoughts it's important to notice the way we speak to ourselves; the tone of voice conveys our feelings behind the words we use and our feelings are 'tagged' with memory accompanied by packages of electrical, chemical and hormonal information whizzing about our bloodstreams.

What does the tone of your inner voice sound like?
Notice how often you use a tone of voice that's gentle or soothing, or calm and collected, or optimistic, making you feel good about yourself.
Observe how often you use a tone of voice that is sarcastic, moaning or angry, pessimistic or resigned, or a tone of voice that undermines your self-esteem.

Once we've identified what we say to ourselves and how we say it, we can begin to delete the programmes and text running in our minds that don't bear fruit. Delete the words, phrases or tone of voice that are counter-productive to us feeling good in ourselves and undermine our achievements and happiness. Start clearing up bit by bit; put the old stuff in the trash, start installing a new version of 'Vocabulary for Happiness' and upgrade to an upbeat and encouraging soundtrack.

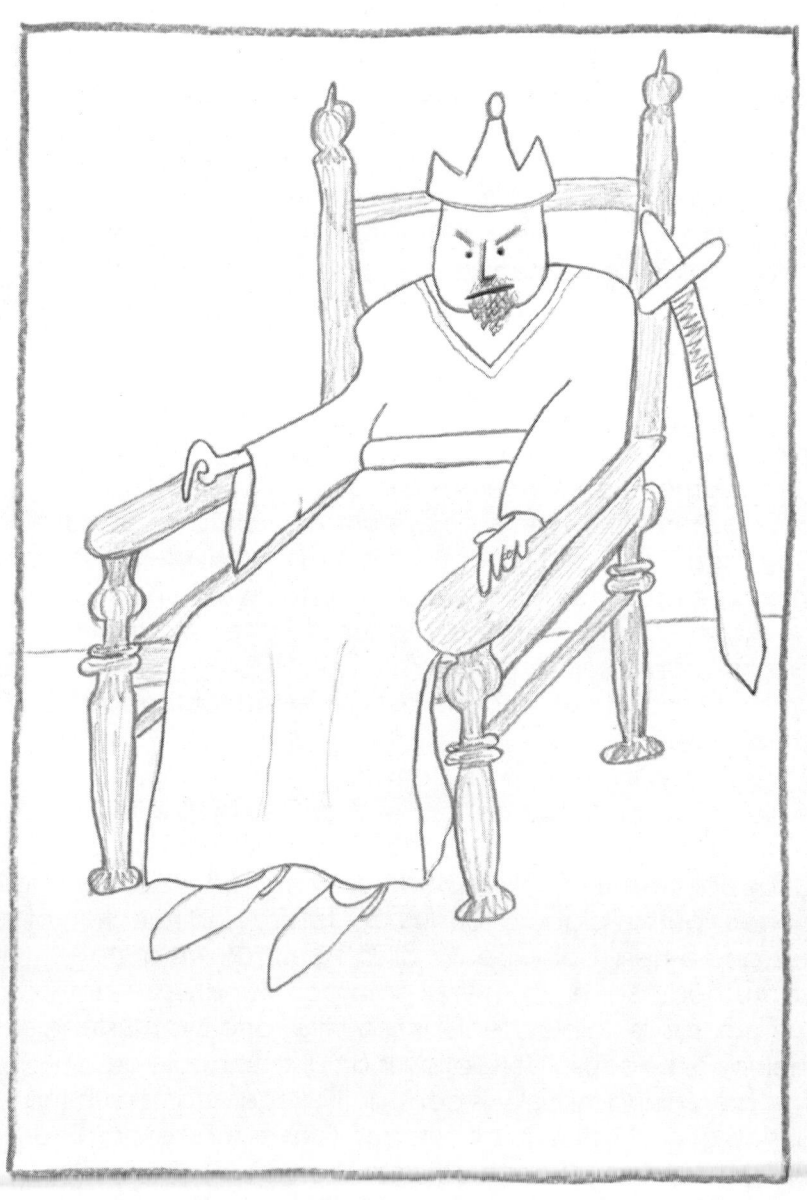

KING CRITICAL

We've all got a programme running in our minds called 'King Critical'. It's probably a familiar voice that we hear too often and sometimes it can seem as if King Critical has taken up permanent residence on our shoulders.

Sometimes it can seem as if King Critical has metamorphosed into an entire Judge and Jury of twenty people all in agreement on the verdicts of: 'Guilty', 'Not Good Enough' and 'Work Harder'. After which, we get sent down for bad behaviour and we're not allowed to play with happiness.

We imprison ourselves with the voice of King Critical. The volume of this voice has become much louder during our current times where we are under so much pressure to be successful. Society's equation of $H = LWH + M$ puts immense pressure on us to look a certain way and strive to be like the photo-shopped images of perfection that we are constantly surrounded by.

Perfection is a tall order and best reserved for the saints amongst us.

If we set our targets up for perfection, if we think and believe we should be perfect or that the people in our lives should be perfect, we are setting ourselves up for a big disappointment. Thinking and believing that we should be perfect puts in place a formula for failure and rejection of self and others, because we aren't perfect, we're not airbrushed. We might as well embrace ourselves exactly as we are, faults, foibles, fears, insecurities and inconsistencies included, because perfection is an unattainable illusion.

The advertising industry is constantly deceiving us with the illusion of perfection in all its forms. Clever marketing distorts our perceptions of each other and ourselves, persuades us to keep buying into these belief systems and keep buying their products. There are quite a few 'fat-cats' out there who have made a lot of money from our insecurities caused by the voice of King Critical.

Changing our targets of perfection to more realistic, achievable and attainable goals, while using our intent to do the best we can, re-directs a big chunk of our thinking; King Critical will be persuaded to put his sword back in its sheath and will be prepared to install new terms and agreements along with the latest version of Vocab for Happiness.

Using our intention to 'do the best we can' is a way to gradually change some of the un-helpful networks of thought in our minds.

It's important to bear in mind that 'our best' can vary from day to day, just like British Summertime; some days we are full of sunshine, motivation and focus and are able to achieve a great deal, whereas on other days we're a bit damp around the edges and although we might not achieve as much, we're still doing 'our best' with the energy we have available.

CHANGE THE TEXT

The words and text of our internal dialogue have a direct and lasting impact on how we feel about ourselves.

A simple example of a phrase that we've all said to ourselves many times, probably hundreds of times is "I should have". I should have done something, said something, been something... the list is endless.

An excerpt from my own internal ramblings goes along the lines of:

'I should have cut the grass last Tuesday before it rained. Yep, I knew I should have cut the grass. Now it's doubled in size, soaking wet and loads more work to do; it'll take much longer, won't cut properly and heavier to empty the box. Feck, loads more work. I knew I should have cut the grass last week. I should have washed the car last week as well. And I should have fixed the hoover ages ago...'

A sentence that begins with "I should have...' is a negative sentence. It's limited, it's final and in the past tense. Even though the emphasis is slight, the word 'should' is none-the-less a 'King Criticism' doesn't make us feel that good about ourselves. We're telling ourselves off.
Just by changing 'should have' to 'could' turns the sentence from a negative, limited and final statement to a positive statement that is open and full of possibility:
"I could cut the grass tomorrow... I could wash the car on Saturday... I could fix the hoover now..."
This is a simple example of changing only one word but the dynamic changes significantly; the tone of voice changes, the implications also change, and the underlying message that we feed back to our subconscious minds has altered into something positive and life affirming.

OMISSIONS
We communicate with each other using words, tone of voice and body language. In conversation it isn't possible to include absolutely every detail or describe the entire context of our thought processes because it would take far too long and would be exceptionally boring all round. So we leave stuff out. We leave out loads of stuff.
We omit much of what we're actually thinking, much of the historical context (which could go back years and have several layers of 'what happened when and who said what to whom'), and we leave out stuff we don't want others to know or because we want to make a certain impression.
We omit much of what has helped to form the thoughts, beliefs and values we have around whatever it is we are discussing. Because we omit so much when we're talking, other people fill the gaps in. However, other people don't

fill the gaps in with what we are thinking, they fill the gaps in with their own thoughts, beliefs, values, perceptions and associations with whatever the topic of discussion is. We all do it, and we can't do anything other than fill the gaps in with our own thoughts.

KANGAROO JUMPING
Filling the gaps in is where misunderstanding can often wedge its hefty foot firmly in the door leaving a muddy trail of erroneous assumptions, wrong conclusions, failed expectations, disappointment, upset and arguments. We're probably all capable of taking a big jump to the wrong conclusion and making assumptions about something. When we're in the act of 'Kangaroo Jumping' we could remind ourselves to ask for more information and clarification and double check we're camping at the same billabong.
Kangaroo Jumping goes both ways in conversation, either when we're listening or talking. In order to avoid a trail of muddy footprints, we could activate our Vocab for Happiness, choose words carefully, aim for clarity of speech and take the courage to be entirely honest.

THE CUPBOARD OF DOOM
Spring-cleaning the thoughts, beliefs and values in our Personal Guide Book to Life is essential for improved happiness and can be undertaken at any time of year. It's a bit like sorting out the cupboard under the stairs that's jam-packed full of stuff, which my daughters refer to as 'The Cupboard of Doom'; it's got thoughts in there that we use, thoughts we no longer use or don't really need, thoughts we've been given but don't want, thoughts that don't work and need fixing, thoughts we don't know what to do with and thoughts we've completely forgotten we had.

Just like spring-cleaning the actual cupboard under the stairs, it can be so jammed packed with stuff that we feel overwhelmed and don't know where to start.

In the first instance, I advise putting on kid gloves or some cotton dusting gloves and to proceed with an attitude of gentle determination. It takes a wee while for the 'Cupboard of Doom' to fill up with stuff so it may take a wee while to sift through its contents and delete any limiting belief programmes that are operative.
Next, we need to ask King Critical to go out for a walk; make a packed lunch, a flask of tea and borrow the neighbour's dog to ensure it's a very long walk.

It's possible that when we open the door to The Cupboard of Doom loads of stuff falls out and lands at our feet, so we have to start by picking that lot up first. Take hold of one thought at a time and decide if it needs to be re-cycled, or up-cycled with different words and mended with care, or if it needs to be put in the 'out box' and ditched completely via the compost, fire or landfill.
Thoughts generally have a trail of other thoughts attached and as we sift through the contents lurking in our Cupboard of Doom, more deeply buried thoughts or limiting beliefs may make their way to the surface of our consciousness. Gradually let go of the thoughts that no longer serve purpose or that don't hold vitality and goodness.

THOUGHT, FEELINGS & BODY CONNECTIONS
Words matter. Words become thoughts and thoughts become feelings. The thoughts of our internal dialogues are directly linked to our emotions and behaviour and the type of chemicals whizzing about our bodies. Taking charge of what we say to ourselves has a beneficial impact on our emotions, our sense of wellbeing and our sense of

happiness. This in turn impacts on our external behaviour and how we relate to others and the world around us. Because of this intrinsic connection between our words, thoughts, feelings and the chemicals released into our bloodstreams, we need to wise-up with our words.

Our emotions and ability to distinguish between and recognise different emotions in ourselves help to create our values. The quality of our internal dialogue, together with the associated thoughts and feelings, help to create the beliefs we have about ourselves and the world around us - the very stuff our individual Guide Books to Life are crafted from. Thoughts affect our emotional state and consequently our physical, external behaviour, what we do and how we do it.

Our body language, how we behave and act, how we sit, stand, how we walk, right down to the tiniest of expressions on our face, physically affect the quality of our thinking and consequently the quality of our emotional state and what types of chemicals are whizzing round our bloodstreams. Because there is an inseparable connection between these parts of ourselves we can catalyse ourselves into a fresh state of being whenever we want. Actively changing any one of these parts, our thoughts, or our feelings or our behaviour/body language, will have a direct and immediate effect on the other aspects of ourselves.

It's not so easy to switch our feelings off when we feel grim. We have to think or do something differently in order for our feelings to shift into something else, something better. When we actively change our thoughts or behaviour/body language, our feelings automatically begin to change too. The signals and chemical messages from our brains change as we think differently or if we physically do something different with our bodies.

We know how it feels to be sat at a computer for too long; it's hard to stay focused, brain fog, backache, neck ache, eyeballs cease to function staring blankly at the screen in a trance of stupefied non-activity, so we take a break. We get a drink, maybe go for a quick walk, grab a bite to eat, bit of fresh air and we begin to think and feel differently; a little bit of movement is enough to change our state of being.

Although this is a small example of how a change in physical behaviour can affect our thoughts and feelings, the principle that we can actively change how we feel is the same for any situation.
It may be that a quick break from the office isn't sufficient to change how we really feel about our job. Changing behaviour may mean altering the overall way we approach our jobs, re-thinking our own attitude, re-thinking our own limiting beliefs and the text of the words we use, re-thinking how we relate to colleagues, or it may mean a radical change of the job itself.

ANCHOR
We're designed with flexi-wiring to enable us to think differently and create new pathways of thought.
If we get stuck with old thoughts re-emerging, we can remind ourselves to Think Think, shift our attention and choose more constructive words and thoughts from our Options Menus.
If it helps, use the image of a 'stop' sign, or a rock in the path of a stream, or any image that serves as an anchor to remind you to stop, and choose a different route of thought. When an unhelpful thought creeps back in or pops up out of nowhere, decide if it has any real value for you and if not, tell it to feck right off.

We only need to keep the thoughts, beliefs and values that nourish our lovely selves and that cherish this beautiful planet.

Think Think.

CHAPTER THREE

THINK THANKS

He is richest who is content with the least
For content is the wealth of nature.

Socrates

THINK THANKS

Gratitude is extremely powerful for making us feel good and for making our Onesies radiate happiness out into this world.

Expressing gratitude for what we have in our lives and vocalising our appreciation to friends, family and colleagues keeps us right on track for having a happy life. There's a host of material currently available about gratitude, hundreds of books, audio books, apps and personal testimonies; gratitude is making a resurgence, and rightly so.

Gratitude is a vital element in the current transformation of humanity in embracing our true, happy nature. The loving force of gratitude is just as essential for us humans surviving on Earth as is the force of gravity.

Thinking, feeling and expressing gratitude sets in motion a pattern of positive pathways in our brains which make it easier for positive thoughts to keep coming, a bit like opening the lock gates on a canal. Opening the gates to gratitude allows for the flow of good, happy feelings to course through our brains, our hearts, our bodies and our Onesies. Grateful Onesies make for good relationships, strengthen the essential connections of our friendships and put us on a solid footing for encountering and meeting new people.

Gratitude comes completely free of charge and because we've got flexi-wiring we can train our brains to 'Think Thanks' more often.

FOCUS
'Thinking Thanks' significantly influences the way our brains do their thinking by tapping into positive patterns and associations when storing and retrieving information. If we

stay focused on what's good in our lives, feel and express gratitude, this state of being becomes more constant and improved levels of happiness become more consistent. The more we value what we have and express our gratitude, the easier it becomes for our brains to make positive associations with whatever else we encounter in our daily lives.

The state of being that accompanies gratitude brings with it a resilience that enables us to hold onto a healthy perspective within the demands of living in the 21st Century. The more we train our brains to Think Thanks, the easier it becomes to remain in a state of wellbeing that enables happiness to permeate right through to our bones.

GRATITUDE RESEARCH

There is growing interest in more research into the effects of gratitude on our levels of happiness. Research already undertaken (1) shows that actively cultivating the quality of gratitude really does impact on how much happiness we experience.

In a study in America, participants were divided into three groups and asked to keep a daily journal about events in their lives for a ten-week period. The first group were asked to write about the things they were grateful for, the second group wrote about things that annoyed or irritated them, and the third group wrote about events that were either positive or negative. At the end of the study, the gratitude group reported feeling happier with their lives and more optimistic; they also took more physical exercise and had fewer visits to the doctor than the other two groups.

GRATITUDE JOURNAL

Writing a gratitude journal on a daily basis is a guaranteed way to improve our happiness. Actively training our thoughts to focus on the good stuff generates more good thoughts. All thoughts come with chemical packages attached, so the more gratitude thoughts we have and express, the more often our bodies are fuelled with the chemicals of goodness; keeping us healthy, happy and connected.

Writing a daily journal might seem a bit daunting and yet another thing to add to the list of 'stuff to do', but starting with a small, achievable target like writing for five minutes, or writing a list of five things that we're grateful for, gets our brains thinking in the right direction.

Writing is a really good way to actively remind ourselves of how much in life there is to be grateful for and feel happy about. Writing in a journal gives us an anchor to return to and re-read when we're feeling off kilter. Writing one daily reason to be grateful means that by the end of a year we'll have 365 reasons to be cheerful.

If we were to write five expressions of gratitude every day, we'd end up with 1,825 reasons to be cheerful.

LETTER OF THANKS

In another study investigating a range of positive interventions, participants were asked to write a letter of gratitude to someone from their past, thanking them for an act of kindness and to deliver the letter in person. Engaging with this task brought the participants an immediate increase in their scores of happiness with long-lasting benefits. Writing a letter of thanks proved to be the most effective exercise of all the study interventions at improving levels of happiness.

How to write a thank-you letter was part of our English curriculum when I was at school, back in the days when we did actually write letters, by hand and on a piece of paper. The advent of emails has made letter writing something of a lost art. We write differently in emails, often opting for the

To
Mrs Cracker
Miss Robertson
8 Mrs Curzon

I'd like to thank you all
for being brilliant and
inspiring teachers throughout
my school years at St. Mary's.
Thank you for appreciating
me just the way I was.

With love from the
bottom of my Onesy

Tami Peirson

p.s. I haven't changed.

shortest reply possible. Writing a letter by hand has a certain personal quality that emails can never replicate. Writing a letter of gratitude by hand brings with it an integrity that showers the reader with a sense of appreciation, something that gets lost in electronic translation.

Writing by hand allows our brains to slow down a bit and think slightly differently; away from the pressure of the inbox and the glare of the screen, we can approach the task of writing in a more relaxed state that engenders the true expression of our heart-felt thanks.

We've probably all got at least one person from our past who we could write to expressing our gratitude for however they may have helped us. If you're worried that your handwriting isn't up to scratch, no worries, you can read it out to them and score extra on the happiness factor. It might not be possible to deliver the letter by hand, but I reckon both the recipients and our selves would still score a fair bit of happy if we sent our words of gratitude via the Royal Mail.

WAKE UP WITH THANKS

Starting the day with thinking Thanks gets our brains in a good mood before we get out of bed. For a start, if we've woken up we're still breathing, we've still got a body and we're still alive. Obviously we can't stay in bed listing everything there is to be grateful for or we would just be in bed forever; the list is endless. Thinking Thanks on waking really does get the ball rolling in the right direction for a happy day. Thinking Thanks is a way to enhance every day, all day. It reminds us to keep both hands firmly on the steering wheel and take charge of the not so good days before they go wonky and start heading off down the wrong road.

SNAPSHOT

If you could take a snapshot of your life in general, what would it look like? Here are a few questions to prompt your brain cells:

How often do you express your gratitude?

Approximately how much of what you think and communicate is focused on what is good in your life, what you have got, what you do like, what you do enjoy?

About how much of what you think focuses on what isn't good in your life, what you haven't got, what you don't like, what you don't enjoy?

How often do you take people, things, food or experience for granted?

If all your thoughts and communications add up to a total value of ten, how much would you score for gratitude and how much would you score for attitude?

CHANGE PERSPECTIVE

It's a big leap to turn things around in our minds and change the way we experiencing something but our flexi-wiring is designed so we can do exactly that. If we keep looking at ourselves, another person or situation from exactly the same place, it will continue to look just the same. Once we start to look from a different angle, what we're looking at also changes and we can view, review and evaluate what is and what isn't working. Taking a bit of time to evaluate our lives from a slightly removed perspective gives us a chance to fully assess what we could do differently. It's a combined state of 'thinking outside the box' and 'thinking outside the head'. From here it's easier to see how our attitudes impact on our lives us and we can identify what resources or skills we might need in order to approach things differently.

Taking a step back in this way also gives us the chance to see things from another person's perspective. Other people's moccasins are extremely good at diffusing an awkward 'attitude situation' back to a place of understanding and gratitude.

It may be that some parts of our lives may not work for us at all and may need radical change, a different job, a different location to live, the end of a relationship or a big change in the way we look after ourselves. These radical changes can be very stressful transitions and this is when a mountain to view things from can be extremely useful. We need to check and double check we've tried different approach roads to 'what isn't working' before deciding to go radical; before quitting the job, moving house or filing for divorce. If we don't check-in completely honestly with ourselves the danger is that the 'isn't working stuff' just jumps in the suitcase and comes along anyway.

If we have an attitude towards someone or something we need to identify exactly what's wrong or what isn't working in order to make necessary changes for what we do want. If we don't stop and look from a wee bit of a distance, we just carry on in the same old way; nothing changes and nothing improves. Our brains stay in the same thought loops and patterns, we use the same manner of communicating and behaving, we invest the same amount of gratitude and the same amount of attitude into our lives.

Once we've identified the 'not working stuff' we can shift our focus to do it differently or do something else.

THE PING-PONG EFFECT

Gratitude plays a very important role in maintaining good relationships. We all want to be appreciated and a sense of appreciation often comes hand in hand with healthy self-esteem and a power boost of motivation. Gratitude has a sort of 'Ping-Pong' quality that can keep bouncing about all over the place, filling us, and others with appreciation. Gratitude plays a lead role in our intimate relationships, but we don't give this character enough words to say.

Sometimes the 'Attitude Character' takes control; our Pointy Fingers twitch away and won't let Gratitude get a word in.

The King Critical voice of our inner dialogue likes to project our standards, expectations and attitudes onto others and if we're not careful, our intimate relationships can descend into a battle between our inner Kings. All our relationships improve when we learn how to temper the voice of these Kings with gratitude.

A study investigating how gratitude impacts on couples in relationships found that expressing this quality more often significantly improved their relationships. The participants who actively expressed their gratitude found that they felt and thought more positively about their partner, and their relationship.

These participants also found greater ease when voicing any concerns about their relationship.

Investing daily in gratitude and rethinking the focus of our communication when relating with family, friends and colleagues, goes a long way to ensuring that our lives and the lives of those we know, are fuelled with happiness. We can train our brains to Think Thanks more often and cultivate the ability to express our gratitude more readily.

FORTUNATE

For most of us, we're fortunate to be living in the UK. We live in a democracy where we have the opportunity to vote for an elected government. Although the system itself could do with re-thinking, improving and representing its people more fairly, we do choose our government.

Most of us do have homes to live in, a bed to sleep in, water, heating and the means to cook and wash with. Most of us have enough food, friends, family, and the freedom to aim for what we want from life.

We have the National Health Service, schools and colleges, public transport, good roads, libraries, unrestricted access to the Internet, good infrastructure and good services. We can wear what we want, go where we want and have endless opportunities to enjoy ourselves, be entertained or go on holiday.

How much of this are we consciously grateful for every day? How much do we take for granted?

GRATITUDE DOESN'T MAKE MONEY

Our society doesn't direct us towards being grateful because gratitude doesn't make any financial profit for the various industries that produce all the stuff for us to consume. Society equates spending power with success and happiness and encourages us to keep spending, even if we haven't got any money and even if it's to the detriment of the planet. This, together with advertising on TV and in the media, directs us into taking much for granted. We're bombarded by adverts telling us we can Buy Happiness, persuading us to buy new products and devaluing the items we've already got, which then get discarded or shoved in the cupboard under the stairs. Consumerism has moulded us into an ungrateful, throwaway society.

ADRIFT

Most of us in the UK have our basic needs met, but something in our culture and society has gone very adrift because there are still many homeless people and families with children living in poverty. The UK falls just behind fifth place in the world for income equality, the biggest divide between rich and poor, yet we are one of the wealthiest nations in the world.

In order not to overwhelm you with the facts depicting the enormous quantity of paper, wood, plastic, textiles, furniture, electrical, electronic and mountains of other goods that we consume and throw away in the UK, this next part focuses solely on one resource that we all need: food.
Looking at just one of these categories is shocking enough. The facts and figures surrounding our consumption and waste of food in the UK highlight our current attitude and give a clear indication of how far we have drifted from gratitude.

WHAT A WASTE

Food waste is a huge global problem with 1.3 billion tonnes of food thrown away every year. One third of all crops grown don't even make it to our plates.

In the UK we currently throw away 15 million tonnes of food a year.

The blunt reality is that we've got a national attitude problem. We buy more food than we can eat, we cook too much, we forget to eat the leftovers or just don't fancy them, and together we throw millions and millions of tonnes of food away every year.
Manipulative marketing has turned us into a wasteful culture not a grateful one.

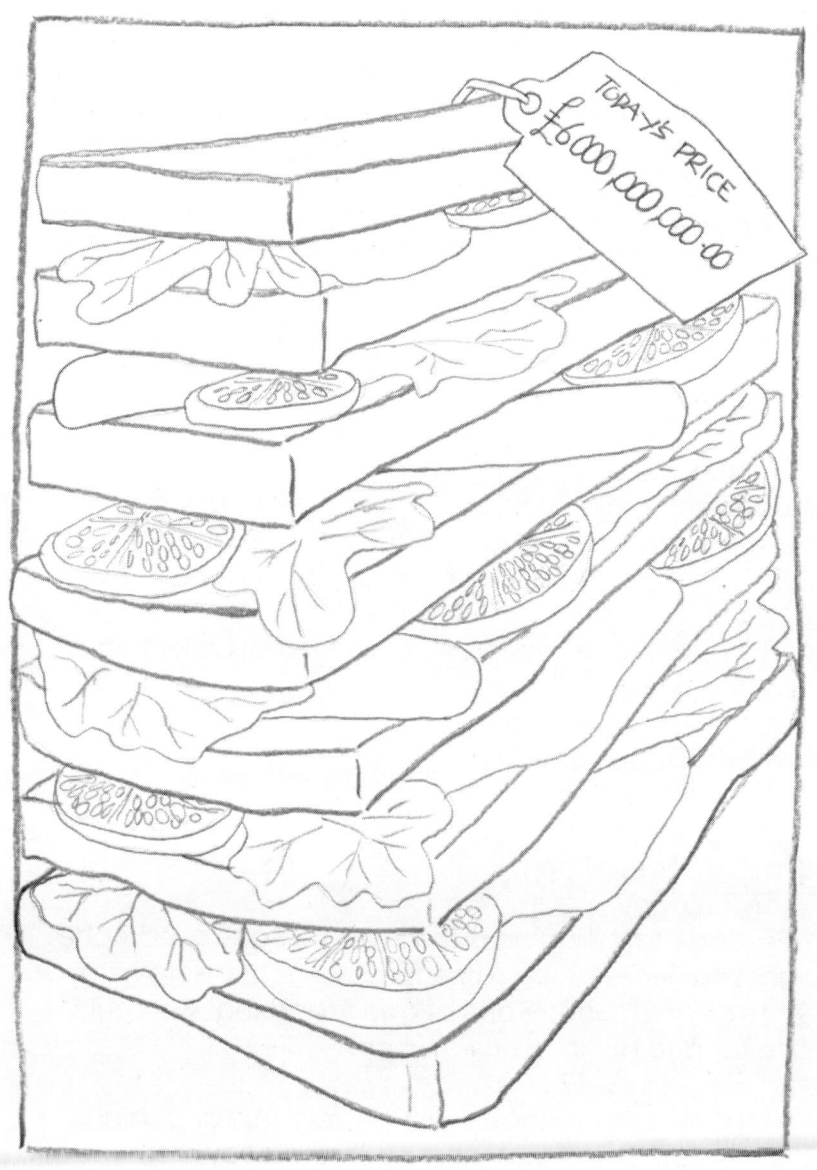

TODAY'S PRICE
£6,000,000,000.00

73

The facts around our food waste in the UK are truly shocking. This next chunk is based on extracts from information on WRAP's website - Waste Resources Action Programme, who are a charitable organisation dedicated to 'accelerate the move to a sustainable resource-efficient economy.' (3)

Food waste in the UK amounts to 15 million tonnes a year. Almost half of this food waste comes from our homes. That's a cost of £12.5 billion pounds on food that just gets thrown away.
We throw away 7 million tonnes of food and drink from our homes and more than half of this we could have eaten, which is equivalent to us collectively putting six billion pounds worth of food in the bin.
Billion is such a big number, we don't relate to it that well, so I'm going to say it again: collectively we put £6000,000 000.00 worth of food in the bin that we could have eaten.

Every day in the UK we throw away the equivalent of:
5.8 million whole potatoes
1.4 million whole bananas
1.5 million whole tomatoes
24 million slices of bread
1.5 million sausages
1.9 million slices of ham
1.1 million eggs
We spend £6.5 billion on buying ready-made sandwiches every year while at the same time we're throwing away over 6 billion pounds worth of good food that we could have used to make our own lunch.

The foods we waste most are fresh veg, fruit and salad, drinks, bread and cakes, and we throw away more food from our homes than we do packaging.

If we stop throwing away food that we could have eaten, not only will we save ourselves a huge amount of money, the benefit to the planet will be the equivalent to taking one in every four cars off the road. The waste of good food and drink is associated with 4% of the UK's total 'water footprint'.

PROGRESS

The good news: between 2007 and 2012 avoidable food waste was reduced by 21% - over 1 million tonnes, and this amount of food would fill 23 million wheelie bins - that's the big wheelie bins. Since the launch of Wrap's consumer campaign – 'Love Food Hate Waste' encouraging us as individuals and families to think differently about the amount of food we buy, the amount of food we prepare and cook, and the amount of food we throw away, consumers have saved £13 billion by not buying food that would otherwise have gone to waste.

Although the statistics are very shocking, thanks to organisations like Wrap attitudes are changing fast.
In May 2015 the French Government made it illegal for their supermarkets (who already donate 20 times more food to charity than the UK supermarkets) to throw away food that could have been eaten. They are now required by law to donate edible food to charities or face heavy fines or imprisonment. (4)
There is pressure here in the UK for our government to do the same. As we currently waste the largest amount of food in the whole of Europe, the implementation of a similar law here would help to change our collective attitude. (5)
All major supermarkets in the UK are already involved with charities to pass on unwanted edible food, to humans and for use as animal feed. However, at the time of writing, only Sainsbury's has a nationwide programme to distribute their

leftover food to charity. Tesco has recently teamed up with 'Fare Share' to run a new scheme where they donate food to local charities a few times each week in 10 stores in the UK, a scheme already tested at 100 stores in Ireland. (6) Things are changing fast, and once the pilot schemes have proven to be successful, similar schemes will start popping up at lots of different supermarkets all over the country.

SUBLIMINAL MESSAGES
Supermarkets redirecting food that would otherwise have gone to waste is great, however the fact remains that most of the food waste comes from our homes. Nowadays, we buy too much, we cook too much and we eat too much grub as well. Which isn't surprising if we take into account the extent to which our shopping habits are being manipulated by advertising and the supermarkets themselves. It's easy to buy stuff that isn't on the shopping list and it's hard to resist a bargain. We are constantly being tempted to buy lots of tasty foods, a huge variety of meals as well as savouries, snacks, puddings, treats, delicacies, pretty much anything we fancy. Our senses are constantly being tempted and persuaded in every which way: the visual imagery on the advertising and packaging show at a glance how good the food looks, the tempting smells of the delicatessen and the bakery are on a direct waft to our brains eliciting hunger responses in our bellies, plus the music or the 'what's on offer' announcements affecting our mood or influencing our choices; everything plays on our senses to slow us down and persuade us to buy more food. Although it's illegal to use subliminal editing in the UK's advertising or music, we are still being manipulated in a

similar way with images, sounds and smells on a direct line to our subconscious minds, influencing us to buy more grub. Consequently we spend more, which puts more profit in someone else's purse, and we eat more, which puts more weight on our bodies.

WE'RE EATING TOO MUCH GRUB

In an article called 'The Truth About Obesity: 10 shocking facts you need to know (7) Sarah Boseley describes how we have an unfair fight on our hands. The food industry is huge and together they spend over a billion pounds every year to ensure their marketing strategies keep us buying their products. Our government is currently spending fourteen million pounds a year to fund the anti-obesity campaign 'Change 4 Life', which is a lot of money but if I've got my maths right, the food industry spends 71 times that amount persuading us otherwise.

Displays at the end of aisles particularly grab our attention as these shelves account for nearly a third of all sales in the supermarket. Companies pay extra to have their produce on these magic shelves and we end up with stuff in the trolley that we don't really need, wasn't on the list, and is invariably full of sugar.

The facts around obesity are indeed shocking. Nearly two thirds of the UK population is overweight or obese, 67% of men and 57% of women and, sadly, a third of all our children. Obesity is defined as being overweight to the extent that it is dangerous to our health with a BMI, Body Mass Index, of 30 or above. Obesity is shattering the lives of many people and often brings with it heart diseases, strokes and diabetes. Society's current attitude towards how much we eat is devastating millions of lives and overwhelming the NHS.

DUPED

Although we seemingly have free choice in what food to buy, it's not entirely our fault that many of us are overweight and struggling with our health. Persuasive advertising and marketing tactics have been established for many decades and our eating habits have been influenced and slowly manipulated. Gradually, we've been coerced and duped into buying more processed food and eating more sugar. Slowly but surely we've become used to more people being overweight and our collective notion of what a normal weight should be has significantly changed. It's changed to such an extent that it's become the norm to be overweight.

In the article there is a photograph taken in the 1950's of a group of boys on the beach, running and probably playing football. They are all wearing shorts so their chests are visible and their ribs are visible too. By today's altered standards of what is normal the children in the photograph look incredibly skinny and under-fed, but they are in fact a normal, healthy weight.

THE LONG VIEW
Much has happened since the 1950's causing a substantial change in our eating habits. In the last few decades our home and working lives have changed along with our beliefs, values and attitudes towards food, what we eat and they way we eat.

Television took off in the 1960's heralding the beginning of a more sedentary life for us all. This came with the start of advertising directly in our homes using a very powerful mix of media, visual imagery with persuasive words and music that has a significant impact on our subconscious minds. The brain's job is to store information with associations, and that's exactly what advertising is, information with associations. Before long, we believe the information and associations that the adverts keep telling and showing us, and all this can end up in our Personalised Guide Book to Life, becoming part of our identity and part of our attitude in life.

The 1970's put more cars on the road and kick-started commuting by car, the beginning of the school run and less exercise for children and adults. There were more and more gadgets to make our lives easier and out-of-town convenience stores. It was the beginning of ready-made meals and eating in front of the telly.

There's been a gradual change in the number of working mothers and with rising costs for everything many families have both partners or both parents, out at work all day. With less time to shop, prepare and cook food we've increasingly opted for more and more convenience foods; it's easy, it's quick and we don't have to think. I've succumbed to many a ready-meal because it's so easy, with no thinking involved.
If we eat in front of the telly we're sort of one step removed and not fully connected to the most important form of nourishment that sustains us and keeps us alive. We're not focused on our food because we're focusing on whatever's on the telly. We don't fully notice or acknowledge the food that keeps us alive. We chomp away, more engrossed in the telly than consciously savouring each tasty mouthful with a grateful heart and a happy belly.

The rise of obesity in the UK began during the 1980's when Mrs Thatcher forged ahead in determined capitalistic shoes and set the country off in pursuit of the illusory American dream: we can have it all and we can have it now. Sarah Boseley reports that the government at the time didn't register the swiftly rising rates of obesity. In 1982, just a couple of years after Mrs Thatcher took office, MacDonald's relocated their main office to her constituency and called upon her to mark the occasion of opening the building. On the 10th anniversary of being Prime Minister, she returned to congratulate MacDonald's for all the jobs it had generated and the economic success of the business.

However, MacDonald's and other fast food outlets are not such an economic success after all. 35 years down the line of eating burgers, fries, pizza, cakes, chocolate, crisps, fizzy drinks, milkshakes, we find that all these types of processed food have become the norm and are making us very ill. Large and even larger portions are available and accepted as being normal. Eating these types of food as part of our every day diet puts tens of thousands of us in hospital and tens of thousands of us in a coffin, every year.

The cultural acceptance of this dietary lifestyle is devastating people's lives, the cost of which is immeasurable - we've only got one life. It's currently costing the NHS five billion pounds a year to treat those suffering with obesity and other related diseases, and predicted to be fifteen billion pounds annually within a couple of decades.

Not such good economics, Mrs T.

THE PUBLIC HEALTH RESPONSIBILITY DEAL
Following the world's biggest conference on cancer in May 2015, all the newspapers reported on the stark facts: obesity is killing tens of thousands of us every year in the UK and significantly increases the risk of common cancers. Medical opinion expects obesity to be the single largest driver of cancer within a decade. (8)

In an interview with Andrew Marr on the 31st May 2015 (Ref 9), Simon Stevens, the Chief Executive to the NHS, said that the food industry does have a responsibility:

"One in three of our teenagers are drinking high energy, sugary drinks, it's the biggest consumption of fizzy drinks anywhere in Europe.
I do think we are going to need reformulation to take sugar out of foods in the same way that has successfully happened with salt over the last several years. I think responsible retailers and food producers can smell the coffee here. They can see that public attitudes are changing and they are going to need to take action because we can see that if that doesn't happen, then in effect what we are doing is a slow burn of food poisoning through all of this sugar that then goes on to cause cancer, diabetes and heart disease.
That's what we're doing to our kids and we've got to stop it."

That's exactly what's happening; sugar's happening. Sugar does cause 'a slow-burn of food poisoning' and it's been poisoning us for some time. Eating too much sugar is exactly what causes weight gain, obesity, diabetes, heart disease and cancer.

Under the coalition government of 2010 - 2015 the 'Public Health Responsibility Deal' was set up which 'aims to tap into the potential for businesses and other influential organisations to make a significant contribution to improving public health by helping us to create this environment'.
Organisations signing up to the 'Responsibility Deal' commit to taking action *voluntarily* to improve public health through their responsibilities as employers, as well as

through their commercial actions and their community activities.

Organisations can sign up to be either national partners or local partners – making collective pledges towards: alcohol, food, health at work and physical activity.' (10)

It's a good idea and gaining momentum but the food industry is slow to improve public health by changing their products. They can afford to pay for substantial advertising and research that will probably conclude that eating processed, sugar-filled, fattening, liver-rotting, cancer-causing, heart-breaking products are perfectly safe for us to eat. Not safe. It might be okay for our health if we only eat these products now and then, but eating this way every day is costing us our health and costing us the earth. Us lot in the Western World are consuming so much that it's become unsustainable for the planet.

ELEPHANT IN THE GREENHOUSE
We all live in the same, enormous greenhouse that has got just the one ozone-layer roof. Venus used to have an ozone-layer roof, but its electro-magnetic power wasn't sufficient to sustain the ozone against the radioactive glare of the Sun, and Venus got fried. As we know, without our roof intact we will also get fried. We're incredibly lucky that the Earth generates sufficient electro-magnetic power to sustain the ozone layer and protect all life on the planet; but meanwhile we're destroying it from the inside.
We are re-thinking our use of fossil fuels and changing to clean, sustainable energy but it seems as if we're ignoring the elephant in the greenhouse; the major contributing factor to climate change is the production and consumption of meat. Our dependency on eating meat and dairy products has literally changed the face of the

Earth. The massive scale of factory-farming cattle and other livestock, currently covering nearly half of the Earths' total land, produces more polluting chemicals into the atmosphere than the worldwide use of fossil fuels for all planes, cars, trains and buses put together. (11)

Millions and millions of acres of Rainforest and habitats have been cleared, and continue to be cleared in order to grow food for cattle, for us to eat. Our meat-rich diet is one of the main causes of wildlife extinction. We've been led to believe that meat is the best source of protein and that we need to eat it to be fit and healthy. Billions of vegetarians and vegans, plus one more, are living testimony to the fact that we don't need meat to be healthy. We've been led to believe that dairy products are an essential part of our daily diet, but dairy products are quite hard for our human stomachs to digest and for some can cause a sort of 'cling-film effect' in the intestines. If intestines can't do their job properly and absorb all the vital goodness from food, our mental and physical health deteriorates and we're deffo less happy.

Eating is essential to stay alive but it's not essential to eat meat.
What is essential is that we have a planet to live on.

SUGAR
Sugar is another food that has been turned into an everyday commodity. When I went online to look up the recommended daily allowance of sugar, I expected it to take me a couple of minutes and provide a sentence or two. I was wrong; I fell into a sticky, black hole, didn't escape the computer for a couple of weeks and came back with enough scary information to fill the next sobering chapter.

The over-consumption of sugar is devastating for our health, our children, our families, our schools and our society and is a major underlying cause of much unhappiness in our personal lives and the world at large.

ATTITUDE V GRATITUDE

The attitude trail has been promoted as progress and has devastated our planet in a very short amount of time.
The attitude trail is devastating our rainforests, polluting the land, polluting our rivers, seas and the water we drink.
Attitude is polluting our skies and the air we breathe.
It's polluting our hearts and minds.
The attitude trail generates millions and millions of tonnes of wasted food in a world where millions and millions of people die of hunger.
The attitude trail is stripping indigenous people of their right to a natural life and is stripping us of our health and happiness.
On the attitude trail we are destroying ourselves while destroying the planet that keeps us alive.

Gratitude brings a genuine sense of happiness.
The more we Think Thanks and feel grateful the easier it becomes for our brains to make positive associations and for us to hold onto the accompanying feelings of appreciation, belonging and happiness.
Gratitude offers an anchor rock for stormy weather, a stabilising force that assists in keeping the worries and pressures of life in a helpful, healthy perspective.
With gratitude coursing through our hearts and minds, happiness comes to court us and together their dance

radiates through our Onesies and shines a light of loving appreciation into our world.

Gratitude holds the mystery and magic of life and can transform any day in a moment, just as a rainbow brings magic to heavy, cloud-filled skies.

With our flexi-wiring we can re-think our attitudes in life and fill the text in our Guide Books with words of gratitude, appreciation and beauty. The more we Think Thanks, the easier it becomes and sets in motion the Ping Pong Effect, influencing others to experience gratitude.
The more we Think Thanks the more we can influence the Collective Happiness Factor bouncing around the world.

We could write a journal and count our blessings each day. We can take a pit stop whenever we want and fill up with a healthy resilience to the demands of life. We can invest gratitude into our relationships and watch them blossom before our eyes and in our hearts. We could find a bit of time to write a letter of thanks to someone from our past. We could buy less food and eat smaller portions. We could eat the leftovers in the fridge, throw less food away and put less money in the bin. We could re-think our individual dietary-footprint and think twice before we fry up the next Venus burger in the knowledge that if everyone in the world consumed resources like we do in the UK, we would need nearly three planets to sustain us all.

We can summon defiance in the face of manipulative marketing and put gratitude at the top of the shopping list.

Think Thanks.

Saying Grace

When we sit down to eat a meal
Thank the many who make us feel
Full of food, content and happy
As dinner began with the farmer chappy
Who ploughed the field and planted the seeds
And nurtured the crop to serve our needs
Then labourers graft and put in long hours
Picking and packing what becomes ours
Then drivers of lorries transport our grub
And deliver it to the market hub
More labouring done by stackers of shelves
So we can easily help ourselves
Then there's those who sit at the till
While we unload and prepare for the bill
All these people plus the farmer's wife
Helped to put food between fork and knife
Remember too the veg of the land
Sparked to action by life's divine hand
Fuelled by sunshine and nourished with rain
Means we're lucky to have dinner again.

Think Thanks

CHAPTER FOUR

THINK SUGAR? THINK CHEMICAL

Refined sugar isn't food.

It's a pure chemical with the formula:
$C_{12} H_{22} O_{11}$
And is more addictive than cocaine.

THINK SUGAR? THINK CHEMICAL

If sugar were personified, I would call it Lucifer after the angel who fell from grace. Originally, sugarcane was considered to be a sacred plant. Today, sugar has fallen from this high status and is wreaking havoc on the health and happiness of millions of people and bringing a living hell to Earth.

This chapter is a sobering perspective on how the 'Sugar Virus' has over-written the text of our Personalised Guide books to Life and is insidiously re-writing our Happiness Programmes.

Refined sugar isn't food. It's a pure chemical with the formula $C12H22O11$ and is more addictive than cocaine. Eating too much refined sugar devastates the liver and the pancreas, causes insulin resistance and is the main driver behind Type 2 diabetes.

Eating too much refined sugar is also the main driver behind weight gain and obesity. Obesity is linked to many other diseases, including cancer.

Eating too much sugar interferes with our moods, impairs our learning and memory, and is linked to brain disorders like Alzheimer's disease.

PLEASURE PATHWAYS

We often have sweet foods to eat as a treat or as a means to comfort ourselves. Sweet foods have become a comfort in a world where there is much to comfort ourselves about. The pressure of everyday life together with global issues can be overwhelming.

One way we buffer ourselves against this pressure and cheer ourselves up is to have a sweet and sugary treat.

Sugar activates the pleasure pathways in the brain so we experience pleasure when eating sugary foods. It's a moment of bliss, intense bliss sometimes, but the sensations are fleeting. We feel good for a while as the sugar level spikes its way up through our brains and bloodstreams, but then the effect wears off, crashes back down and takes us with it; we're left feeling low in energy, tired, probably irritable and finding it harder to focus.

We aren't necessarily aware of how much sugar we're eating.

Cakes, biscuits, puddings, desserts, treats and chocolates are all quite upfront about being sugar-laden. However, we probably don't bother to find a magnifying glass in order to read the ridiculously small print stating the exact amount of sugar we are about to ask our liver and pancreas to process for us. With our sugar-addled attitude, we don't want to know and we just tuck in. Yum yum.

HIDDEN SUGAR

Other foods, savoury items that we don't associate with being sweet, and pretty much anything processed, ready-made or in a packet, also contain refined sugar. Far more sugar than we realise.

After escaping the black hole, I did go in search of a magnifying glass and was appalled to find that hidden sugars have sneaked into many savoury foods. There's even sugar added to bread, at least a gram per slice. A gram might not sound much, but it swiftly adds up, surreptitiously adding sugar to our bloodstreams, de-wiring our brains, and fattening our livers and our bellies. Refined sugar acts like a virus over-writing the text in our Guide Books with the belief that eating sugar is okay and being overweight is okay too, it's normal.

HOW MANY TEASPOONS A DAY?

In the UK, we eat loads of refined sugar.

We eat loads and loads and loads of sugar.

Over the last few hundred years the average person in the UK has gone from eating virtually no sugar at all, to eating 62.5 kilos, or 138 pounds, of sugar a year. Hard to believe, isn't it?

This equates to 1.2 kilos, or 2.6 pounds, of sugar a week, every week. Translated into teaspoons, it's an astounding 238 teaspoons of sugar weekly, or 34 teaspoons a day. (1) Still hard to believe isn't it?

3 TEASPOONS A DAY

It's shocking that we eat so much sugar and even more so in the light of what scientists say is too much. Scientific investigation has shown that our bodies are designed to assimilate, at the very most, 14 grams of sugar a day. That's 3 teaspoons, absolute tops. The maximum quantity varies depending on a person's height, weight and health; the 14g maximum is aimed at people in good health, with healthy body weight and above average height.

Any more than 14 grams and we're in trouble. And we are in big trouble. Two thirds of our population are overweight because eating so much refined sugar has become normalised. It's become an acceptable part of our culture to eat about 12 times the amount of sugar our bodies can physically cope with, every single day.

CONNED

Because sugar activates the Pleasure Pathways in our brains it's easy to believe that sugary food and drinks bring happiness, so we keep buying sugar-filled products. Advertising takes advantage of this and have conned us into believing slogans like 'A Mars a day helps you work, rest and play'. It doesn't. In the long run, eating high quantities of sugar every day makes us extremely ill.

A standard 54g Mars Bar has 31.2g of sugar, which is equal to 6.5 teaspoons and is more than double the quantity scientists say is safe for our bodies to ingest in one day. With our sugar-addled attitude, a Mars Bar is considered to be a snack to keep us going, regardless of the high sugar content. By the time we've added in all the other sugary stuff, hidden sugars and natural sugars we eat, it becomes easier to see how the average person in the UK consumes 34 teaspoons of this toxic chemical every day.

UK AVERAGE DIET
Here's a quick look at what an average daily diet in the UK may include:

Breakfast
Cereals already have sugar added: 2-3tsp per 30g serving but we still put more on, so that probably adds up to 4tsp. Tea/Coffee. I used to put 2 teaspoons of sugar in my coffee until I fell in the black hole. Let's say the average person has 2 cups of tea/coffee with one teaspoon of sugar.

Breakfast Total = 6 teaspoons of sugar.

Mid-morning snack
Fancy a Twix? A standard bar has 10 teaspoons of sugar. (49.2g)
Mid-morning fizzy drink – A 330ml can of coke has 7.5 teaspoons of sugar. (37.7g)

Mid-Morning Total = 17.5 teaspoons of sugar
Lunch
A sandwich from a shop (even though we could have made lunch from the leftovers in the fridge) will have a few grams of sugar added to the bread, plus other hidden

sugar if we have pickle or sauce, which probably adds up to at least one teaspoon of sugar.

Followed by a yoghurt maybe? How about a 'Muller Corner Health Balance Tropical Crunch'? It has 'Health Balance' in the title, so must be healthy, right? Nope, not healthy, it's got 4 teaspoons of sugar. (21g)

A lunchtime drink – How about a bottle of 'This Juicy Water - Oranges and Lemons'? Sounds healthy. It contains 9 teaspoons of sugar (45g) per bottle.

Lunch Total = 14 teaspoons of sugar

Mid-afternoon snack
Biscuits have about 4.5g of sugar each and they usually come in threes, so that's another 3 teaspoons.
Mid afternoon drink – Tea/Coffee - another teaspoon of sugar.

Mid-afternoon Total = 4 teaspoons of sugar

Dinner
There are naturally occurring sugars in the carbohydrates we eat - pasta, potatoes, rice, vegetables, and then there's pudding.

How about a Co-op Mandarin Fruit Tart? At a glance I can see the red line for fat content and an orange line for sugar content. On closer inspection with magnifying glass in hand, I can see it has 5 teaspoons of sugar per serving (24g) which is described as a 'medium' sugar content.

The day's nearly over, apart from a little snack or drink in the evening and we're at a total of 46.5 teaspoons of sugar.

Many of the items I've put in the list are particularly sugar-rich, so a tad exaggerated but it's a bit easier to understand how the average one of us, myself included, manages to chomp through loads, if not 34 teaspoons, of sugar each and every day.

THE REWARD SYSTEM
Because the pleasure pathways in our brains are activated by sugar- rich foods it's really easy for us to get hooked on the stuff. Sugar activates the same neural pathways that are activated by cocaine, heroin, alcohol and tobacco. (7)
This particular set of neural pathways is referred to as the 'reward system' and when activated sends chemical and electrical messages throughout our brain, releasing dopamine into our bloodstreams.
We like the experience of dopamine in our bloodstreams. If we supply our bloodstreams with more of a substance that triggers the reward system, the dopamine receptors can get quite noisy demanding more and more and eventually we can get hooked on whatever ignites these pathways. Although the effect of sugar on our brains is not as extreme as other addictive substances, sugar nonetheless activates the dopamine receptors in our reward system.
Over-active receptors that have acquired a taste for sugar become very talkative and won't shut up until we feed them more of the sugary stuff. Sugar cravings become more intense as our dopamine receptors become more and more demanding, we lose control and, with our sugar-addled attitude, eat yet more sugar in one form or another. As we become increasingly more tolerant to sugar, we need to eat more in order to activate the reward system and fill our bloodstream with that fleeting, pleasant feeling that dopamine provides. It's no wonder we eat so much of the stuff. (8)

CHANGE OF DIET

Our diet has changed dramatically in the last few decades with the manufacture of processed food and the rise in fast food outlets, and our beliefs, values and attitudes have changed along with our diet. We are encouraged us to eat over-sized portions and sold 'Death by Milkshake' drinks containing 13 teaspoons of sugar in one drink (65g). African people are no longer enslaved to our sugar-cane industry; it is us who are enslaved to sugar. The price at which we now eat sugar in Europe is rising dramatically all the time; tens of thousands of people die in the UK alone from obesity-related diseases.

The rise in Britain's annual per capita sugar consumption over the last few hundred years is very shocking. By 1900 it rose to the extent that the British consumed the most sugar in the whole of Europe. It's possible we still do; currently our children drink the most sugar-laden drinks in the whole of Europe.

Here are the figures of the rise in Britain's annual per capita sugar consumption: (3)

1704 – 4lbs / 1.8kg
1800 – 18lbs / 8kg
1901 – 90lbs / 40kg
2015 – 138lbs / 63kg

On average, we currently consume 1.2kg of sugar a week. That's not far off the amount we consumed in a whole year back in 1704. The maths says that in 1704 we ate just under 5 grams of sugar a day. One large teaspoon tops. Now we eat 34 times that amount, every single day of the year. FFS.

The effect of sugar on people's health didn't go completely unnoticed throughout the years, just unheeded. In 1675, Thomas Willis, physician and a founder member of Britain's Royal Society (2), noticed the sweetness factor in the urine of people with diabetes. He described it as:

"Wonderfully sweet, as if it were imbued with honey or sugar." During 1900 and 1920 there was a huge rise in deaths owing to diabetes, which was noted by Haven Emerson, from Columbia University, as having a direct link to the huge rise in the amount of sugar in people's diets. Forty years on, John Yudkin, a British nutritionist and well-known expert in this field, undertook scientific experiments to demonstrate that high levels of sugar consumption lead directly to higher levels of insulin and fat in the bloodstream, the main drivers behind diabetes and heart disease. However, Yudkin's voice wasn't heard amidst the noise of different scientists claiming that eating too much saturated fat was the culprit causing diabetes and heart disease.

One of the consequences of this dominant line of thinking was that the fat content was removed from many foods, rendering them quite tasteless so more sugar was added instead. Another consequence is that many of us grew up with the belief that fat makes us fat. Good fats do not make us fat. Refined sugar makes us fat and extremely ill. Another 50 years on, 2018 and the truth about sugar is still being deflected by our governments, and by the manufactures of sugary products, and unfortunately we're still chomping away on far too much of the sugary stuff.

NO 'OFF SWITCH'
Back in the day when our genome was being imprinted for our long-term survival, we didn't need an off switch for sugar as we only had access to fruit for a short period

during the year. We could gorge away on seasonal fruit and any excess sugar got converted into glycogen by the liver, and stored to assist the body during leaner winter months.

A piece of fruit comes ready packaged with all the minerals and vitamins our bodies need to assimilate the natural sugar in the fruit.

Unfortunately the processed, refined sugar that we eat today doesn't have any nutritional value and the essential nutrients needed to metabolise sugar are stripped from elsewhere in our bodies.

And without an off switch, the brain doesn't know when it's had enough sugar.

GLUCOSE AND FRUCTOSE

Glucose is the main fuel for our brains and bodies. (4) We need this energy in the tank to stay alive and keep going. Every single cell in every living thing contains glucose and quite remarkably our bodies will produce glucose if we don't get sufficient quantities in our diet.

Fructose is in another category all together. We do not have a physiological need for fructose in the way that our bodies need glucose. Eating fruit is beneficial for our health and comes ready packaged with all the nutrients we need to process the sugar.

However, eating too much sugar extracted from fruit and refined into fructose is a real badass for our brains and bodies. Fructose and high fructose syrup are added to many foods and drinks in much greater quantities than we'd find in a humble apple or banana. For many of our population who eat a typical Western diet with high sugar content, high carb and the wrong fats, the liver is already full up with glycogen. With no more storage space available, the liver has no choice other than to convert the sugars to fat, which is stored elsewhere on our bodies. If the liver is continually given too much sugar to process, it gets extremely pissed off and becomes susceptible to non-alcoholic fatty-liver disease, which leads to many other health issues.

Many food products and drinks have refined sugar and fructose added in extremely high quantities. A concerning factor about fructose is that it doesn't reduce the level of the hunger hormone ghrelin in our bloodstreams to the same extent that glucose does.

If the ghrelin gremlins are still grumbling, we are left feeling unsatisfied and before long we'll be reaching for more of the fructose-filled grub.

INSULIN INSPECTOR

Consuming too much refined sugar of any type plays havoc with our bodies' natural insulin levels. The black hole informed me that insulin is a hormone that has several extremely important jobs to do. It's crucial in the absorption of glucose from our bloodstream into all cells and sends the command, " Burn up glucose Baby Face and save the fat for a rainy day." It might not be those exact words; your 'Insulin Inspector' might speak in Welsh. Whatever language, Insulin Inspectors need to be forceful because too much sugar in the bloodstream is extremely toxic. If our bodies are continually processing high levels of sugar, our Insulin Inspector gets overworked and our bodies begin to become resistant to his commands. This is when the pancreas has to manufacture more insulin to counteract the potential toxic effect of too much sugar in our bloodstreams. If we continue to eat the same sugary, high carb diet, after time the pancreas also gets worn out and can no longer produce enough insulin to lower the blood-sugar levels back down to a safe level. A person with a pancreas that has been worn out in this way is in danger of developing Type 2 Diabetes.

Sugary drinks are the worst. With no substance, no vitamins, minerals, enzymes or proteins, these drinks go straight to the liver and come with an 83% greater risk of developing Type 2 Diabetes. (4)

WORLD HEALTH ORGANISATION RDA FOR SUGAR

In another article in the black hole, (5) Lizzy Parry refers to the research undertaken by scientists at the University College London specifically looking at the effect of sugar on our pancreas. It's these scientists who found that the pancreas can't process any more than 3 teaspoons or 14g of sugar a day.

Unfortunately the advice from the World Health Organisation doesn't reflect what the scientists are advising as a recommended safe level. Currently, the WHO's recommended daily allowance for sugar remains at 'a maximum of 10% of total energy intake (calories) from free or added sugars, with 5% as target.'

These figures, however, do not tally with the scientific advice.

Here's a bit of maths:

14g / 3 teaspoons of sugar is equivalent to 3% of total calories.

50g / 10 teaspoons of sugar is equivalent to 10% of total calories.

25g / 5 teaspoons of sugar is equivalent to 5% of total calories.

Which means that the WHO recommended target @ 25g has 11g more sugar than the advice of the scientists.

If 14 grams is the total maximum daily amount of sugar that our pancreas can process, adding another 10 grams into the equation is equivalent to 72% pancreatic overload.

50g / 10 teaspoons of sugar a day is more than three times the amount recommended by scientists, and is equivalent to 260% pancreatic overload.

170g / 34 teaspoons of sugar that the average one of us consumes on a daily basis in the UK, is twelve times the scientific advice and is equivalent to 1,123% pancreatic overload.

That's scary.

I think I'll get someone to check my maths.

On the good side, some determined people are on the case and are not going to sit down and keep quiet: 'Action on Sugar' advisor Tam Fry, who is also Patron of the UK's Child Growth Foundation, (Ref 5), overtly challenged the WHO for not changing the RDA of sugar from 10% to 5% and insisted on a specific recommendation for children. Meanwhile the sugar industry is putting huge pressure on the World Health Organisation to keep the guidelines at 10%, claiming that there isn't sufficient evidence to lower the guidelines. The sugar industry lobbyists have even claimed that there isn't sufficient evidence to substantiate the link between refined sugar consumption and tooth decay. Which leaves my eyebrows, and my remaining teeth, hanging in a state of incredulity. Currently, the most frequent reason for our young children being admitted to hospital in the UK is tooth decay.

The bigger picture in the UK and throughout the developed world is a grim and unhappy spectacle; we have a mass over-consumption of refined sugar going on. We don't even come anywhere near the 10% mark, which is 10 teaspoons a day, 260% pancreatic overload. And it's not surprising really; our food is absolutely stuffed with the addictive chemical, so no wonder that so many of us are in trouble with our health.

GOODBYE OLD FRIEND

On escaping the black hole, my first confrontation was with an old friend who had been the best of friends since my early childhood days. This friend was a continuous source of reliable comfort and always willing to respond to my needs. We had great times and would travel everywhere together. Fifty long years of friendship. It was quite a confrontation when I finally came to have words with Chocolate. We were on exceptionally good terms until the black hole incident.

One small bar of Cadbury's Dairy Milk, 36g, has 6 pieces of chocolate. When I was a kid, a bar this size was standard. I don't remember many variations. The bars in the vending machines were a slightly different shape and tasted much better on account of illicit spending of church collection money, and on a Sunday too. By today's standards, 36g is a rather small bar of chocolate amidst the medium bars, large bars and ridiculously large bars.

The small 36g bar contains 20g of sugar and the information on the packaging states that this small bar of choc is 22% of an average daily calorie intake.

This one small bar of chocolate contains more than double the WHO recommended daily allowance of sugar at 10%, more than four times the target of 5% and more than seven times the amount of sugar advised by the scientists at 3%. Just in one small bar of chocolate.

WOOLLY WATER

Flavoured water drinks are replacing fizzy drinks but they still contain a high level of sugar with 13.5g per portion, a whole day's worth of sugar in one drink. This type of drink has become extremely popular with brands like 'Innocent' and 'This Juicy Water', both owned by Coca-Cola, setting the trend and successfully pulling the wool over our eyes.

High consumption of these types of drink carries a very high risk of developing diabetes, Hashtag 83%, and our kids are guzzling down hundreds of millions of litres of the stuff.

MORE ADDICTIVE THAN COCAINE
One of the main reasons why no developed country is anywhere near reaching the World Health Organisation's guideline of 10% maximum RDA of sugar is that sugar is so addictive.
I found it remarkably easy to lose control when Chocolate and I were still mates. I would even get up in the middle of the night sometimes just to check in and show my appreciation. Some people sleepwalk during the night; I used to Choco-Walk.
I did the cold turkey thing in the end and re-wrote some of the text in my Personalised Guide Book to Life. I changed the words "I love chocolate" to "I don't eat chocolate." And because of the stark truth in the black hole, it's still working.

Scientists from France (who don't have any knowledge of my Choco-Walking) believe that we really can become addicted to sugar in the same way we can get addicted to other harmful substances. Scientists investigating the addictive nature of sugar found that rats would choose sugar over and above cocaine, even if they were already addicted to cocaine. This suggests that sugar is more addictive than cocaine, although it probably doesn't make us talk nonsense to quite the same extent.
The French scientists think that humans and other mammals are not designed to be able to adapt to the extremely high levels of refined sugar that we are currently eating. Their concerns, published in 2007, are that constantly over-stimulating the dopamine receptors and the reward system

with high levels of sugar leads to a lack of self-control and an addictive pattern can set in.

$C_{12} H_{22} O_{11}$
A sugarcane plant is a complex mix of lots of amazing stuff. (7)
It's made up of fibre, protein, vitamins, minerals, amino acids, enzymes, unsaturated fat, a host of 64 different elements, and of course all the spaces in between these molecules as they wiggle about vibrating with energy. The refinement process that the sugarcane plant undergoes to make the white stuff our dopamine receptors like so much is so very extreme that there is absolutely nothing left afterwards, bar the white powder.
The remaining white powder is a pure chemical with the formula $C_{12}H_{22}O_{11}$. It's a purer chemical than cocaine and has remarkably similar components to cocaine. This chemical is known as sucrose and is a potent mix of Carbon – 12 atoms, Hydrogen – 22 atoms, and Oxygen – 11 atoms. Cocaine is a similar mix of these three elements plus a bit of added nitrogen for the continuous-talking-of-rubbish effect. The formula for cocaine is $C_{17} H_{21} NO_4$. Not a lot in it really. Back in the day, (yonks ago) both were probably only used medicinally. Today, both are substances of abuse and an underlying cause of much unhappiness. Obviously there are quite a lot of differences between sugar and cocaine, not the least of which is that sugar is completely legal. At one point in time, Coca-Cola contained cocaine in its ingredients; it was named for cocaine. When cocaine became illegal,
this chemical substance was removed from the drink Coca-Cola and replaced with a massive dose of a chemical even more addictive: $C_{12}H_{22}O_{11}$; refined sugar.
One 330ml can of Coca-Cola has 35g / 7 teaspoons full of $C_{12}H_{22}O_{11}$.

THIEF

Because the process of refining sugarcane completely removes all the nutritional goodness that our bodies need to metabolise the sugar, our systems have no other choice than to steal the calcium, sodium, vitamins, other minerals and vital enzymes from elsewhere in our bodies in order to assimilate the sugar. Regularly eating too much sugar can seriously deplete the calcium from our bones, leaving them extremely unhappy and highly susceptible to osteoporosis.

ACIDIC

Sugar is acidic and eating it as part of our everyday diet causes our metabolisms to become over-acidic. This forces our bodies to extract more vitamins and minerals from our blood and bones to counteract the acid. If our bodies become so depleted that they can no longer metabolise the sugar, toxic residues get left behind. These toxic wastes get stuck in our bloodstreams, stuck in our nervous systems and stuck in our brains and can result in carbonic poisoning. Perhaps the twelve atoms of carbon bouncing about every molecule of sugar are the culprits here.

STICKY

As we know, sugar is very sticky; it also thickens our blood, making it very sticky too. Blood doesn't like being sticky as it can't do its job properly and flow smoothly along our veins into every nook and cranny via tiny capillaries. Some of these tiny capillaries form the lifeline for our gums and teeth. With sticky blood our teeth and gums are left feeling extremely unhappy, unwell and ravenously hungry.
Us lot in the UK, together with our mates in the USA, are among the largest consumers of sugar in the world and

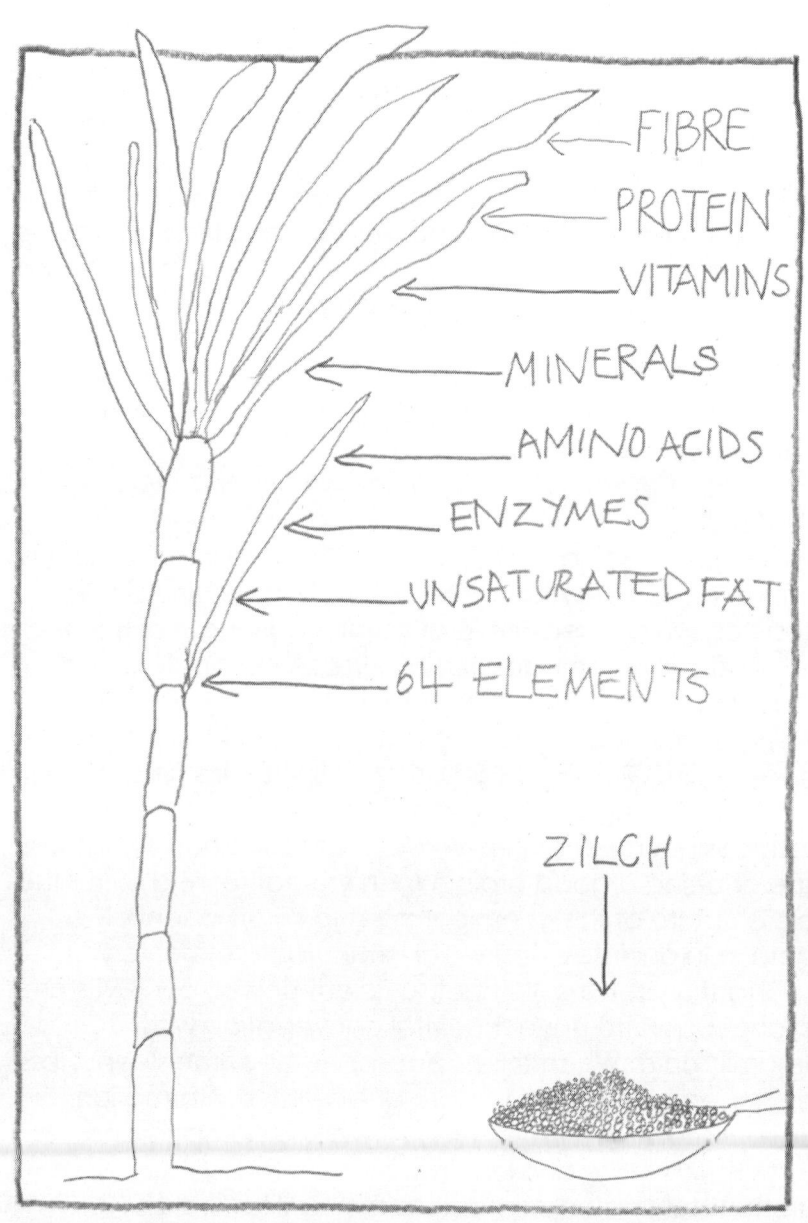

FIBRE

PROTEIN

VITAMINS

MINERALS

AMINO ACIDS

ENZYMES

UNSATURATED FAT

64 ELEMENTS

ZILCH

have the teeth to testify to this status. Or rather, our lack of teeth is testimony to our sugar status.

MORE GRIM SHIT
There's more. We're still in the black hole with our sugar-addicted brain, worn out pancreas, diseased liver, diabetes, heart disease, obesity, carbonic poisoning, depleted bones, rotten teeth and increased risk of cancer. $C_{12}H_{22}O_{11}$ also plays havoc with our intestines, interferes with our moods, impedes our learning and memory skills and, not surprisingly, a high consumption of refined sugar is also linked to brain dysfunctions, including Alzheimer's disease.

INTESTINES
There's no escaping the fact that everything in our bodies is connected to everything else. The state of health of the intestines, which measure in at combined length of 8.5m/28 feet, has a direct impact on how well our brains function. What we eat does affect our thinking.
Glutamic acid is in lots of vegetables and is vital for a happy brain. Glutamic acid and the B vitamins have a special, on-going conversation that somehow magically produces neural enzymes to keep our brains happily on the case of being a good brain. When we eat refined sugar the acidic nature of this chemical starts to kill off essential bacteria in our intestines, which manufacture the very important B vitamins. Without sufficient power, our B Vits are no longer able to have that vital conversation with Glutamic acid. When this happens, we become sleepy, less able to focus, less able to store and retrieve information from our short-term memory, and we find numbers and maths much more challenging.
Unfortunately all food types lose their B Vitamins, as well as all the other vital nutrients, when processed. Which leaves

loads of us who eat processed food wobbling about without enough 'B Vit Power' to sustain the vital conversation that keeps our brains happy and staying on the case. (7)

SEROTONIN

In an article called 'This is what sugar does to your brain' (Huffington Post, 6th May 2015, Ref 9), Carolyn Gregoire refers to Dr Datis Kharrazian, a functional medical expert and author of 'Why Isn't My Brain Working'. This article describes how sugar puts a big spanner in the works, interferes with our moods and is linked to an increased risk of brain diseases. Not so cheery stuff.

In our brains we have a bunch of lively and very talkative chemical messengers, known to scientists as neurotransmitters. Part of their job description is to relay and regulate information and commands, and blast a bit of electro-chemical information to correspond to whatever mood we are experiencing. This lot are totally on the case. As well as a load of other jobs to do, they tell our hearts to keep beating, our lungs to breathe and our bellies to digest. Amongst these Chatty Chappies is serotonin.

Serotonin is everyone's friend because it always wants to play and always puts us in a good mood. Eating sugar gets Serotonin bouncing about our blood and brain cells like nobody's business, but eating sugar regularly and over-activating our Serotonin friends knackers them out completely. It leaves them with no bouncing power whatsoever and before long they don't even want to come out to play. When serotonin levels become this depleted it can lead to symptoms of depression. A continually high blood sugar level is associated with inflammation of the brain, another possible contributing factor to depression.

TYPE 3 DIABETES

Research is also suggesting that our brains are particularly susceptible to damage caused by too much of the chemical $C_{12}H_{22}O_{11}$ in our bloodstreams. There is growing evidence indicating a strong connection between the possibility of developing Alzheimer's disease and eating high quantities of refined sugar in our diet. Research in 2013 connected the main drivers behind diabetes, insulin resistance and high blood sugar levels, with an increased risk of brain disorders such as Alzheimer's disease. Some of those researching this field have described Alzheimer's as 'Type 3 Diabetes', clearly suggesting the link between eating too much refined sugar and the health of our brains.

DIET & COGNITIVE SKILLS
In the article 'UCLA study shows high-fructose diet sabotages learning and memory' (10) Elaine Schmidt reports on a study led by Fernando Gomez-Pinilla, who is a professor of neurosurgery, interactive biology and physiology. The focus of their investigation was on refined fructose that's been added to processed food as a preservative or sweetener.
The study examined the impact of a diet with high levels of fructose on two different groups of rats. Firstly, all the rats were put in a maze to establish their cognitive ability to learn and remember the escape route, before being separated into two groups with different diets. Both groups had high-fructose water to drink but one group also had omega 3 fatty acid and DHA added to their diet. The experiment lasted for six weeks, after which both groups were tested on their ability to recall the route through the maze and escape.

The rats that had added Omegas and DHA in their diet managed to hold onto their brainpower, remember the route through the maze and escaped much faster than the other group. The rats without added DHA displayed poor cognitive ability, showing a decline in the ability of brain synapses to signal properly to each other. Synapses are the bits at the end of all neurons in the brain and they transmit information, a bit like a tin can telephone but without string in the gap. Chemical information gets 'fired' across the small gap so that one synapse can talk to another. It's this chemical conversation between our brain synapses that equip our brains with the ability to learn and remember information and generally do the job of being a brain. If synaptic activity is impaired, it makes it much harder to think clearly and remember information, regardless of what type of mammal the brain belongs to.

Professor Gomez-Pinilla concluded from this study that what we eat really does affect how we think. Regularly eating a diet that's high in fructose changes our brains' functional ability to remember and learn new information. The study also highlighted how including omegas and DHA in our diet significantly helps to counteract the damage caused by fructose. The omegas and the hard to pronounce Decosahexaenoic Acid – DHA - are essential for protecting our synapses against damage and maintaining their ability to keep talking to each other. We have to include foods with Omegas and DHA in our diets, as our bodies do not produce enough of these essential fatty acids to keep our synapses feeling happy and chatty.

Another important factor highlighted in this study, is that the rats without DHA in their diet showed signs of insulin resistance. Insulin has an important role regulating the function of our synapses as well as controlling our blood sugar levels. The Insulin Inspector's job here is to tell our brain cells either to use, or to store, the sugars in our blood so that our brain cells have enough fuel to process our thoughts and emotions. Insulin has the ability permeate through the blood-brain barrier. The blood-brain barrier is a membrane of blood vessels surrounding the brain that's a bit like Fort Knox; it will only allow essential nutrients to enter the hallowed ground of our brains and will block all other unwanted substances. Inspector Insulin knows the security code and can access the inner sanctum of our brains but too much of this hormone can cause confusion to our neurons, triggering chemical reactions that interfere with learning ability and also causes loss of memory.

So along with everything else, this type of sugar makes us stupid.

THANKS

I'm highly grateful for all the articles I found in the black hole, and very grateful to Elaine Schmidt, Professor Gomez-Pinilla and especially all the rats, for this last chunk of information. A recurring question that had been on my mind was "Why have I only got half a brain?" Reading this article gave me all the answers that my doctor couldn't provide.

About 10 years ago, I swapped eating regular refined sugar/$C_{12}H_{22}O_{11}$, for refined fructose, which has the formula $C_6H_{12}O_6$. I changed to 'Fruit Sugar' because I believed it was better for me. The packaging describes the lower Glycaemic Index of fructose, which releases its sugar slowly and doesn't cause blood sugar levels to spike and crash in the same way as regular refined sugar. The chemical fructose, $C_6H_{12}O_6$, does however seriously interfere with brain function and without an off switch my brain just didn't know when it'd had enough. After ten years of a generous daily dose of this poisonous chemical, I was very aware that my brain wasn't working properly but didn't know why. It was a big relief to discover the reason

behind my 'half-a-brain syndrome' and that I could easily do something about it. I think a skull and cross bones would be a more fitting image on the packaging than the pretty picture of a harmless bit of fruit.
And glad to say, thanks to this article, I'm getting my brainpower back.

The information in the black hole provides much to think, and re-think, about sugar. If we want to improve our health, improve our brain function and ultimately improve our individual and collective levels of happiness, we need to re-think sugar.

THE GOOD NEWS
In December 2016 the government published draft proposals (12) to put a tax on sugar-sweetened drinks, to be implemented in April 2018.
There are two levels of taxation, one for soft drinks containing more that 5g of sugar per 100ml, and a higher levy for drinks with more than 8g per 100ml. Fruit juices are exempt from tax although the maximum recommendation per day is no more than 150ml. Because ministers writing this legislation are concerned that teenagers don't get sufficient calcium in their diets, particularly girls, sugary milkshakes and yogurts are also exempted from tax. While a lack of calcium in our diets is a concern for everybody, sugary milkshakes are the worst. A large McDeath Milkshake has 13 teaspoons of sugar and scientific evidence has shown that too much sugar in the blood leeches vital nutrients from our bones and can cause osteoporosis.

The Action on Sugar campaigns are working; the World Health Organisation has recently altered their recommendations on the daily allowance for sugar and currently have a 'strong recommendation' to reduce sugars intake to less than 10% of total daily calories with a further 'conditional recommendation' to reduce sugars intake to less than 5% of total daily calories. This might only be a small shift in the wording of their recommendations, but it is a massive shift in the right direction for many countries to re-think the consumption of sugar.

More encouraging news on the WHO's website is that Tesco are the first to announce that they will 'commit to reducing added sugars by 5% incrementally a year in all its entire sugary soft drink range.' (11) Many brands are following suit and reducing the sugar content in their products.

CHOICE
We can choose to re-think sugar and choose not to eat so much of this toxic chemical. We can wise up and stop buying products that trick us into believing they are healthy because of clever marketing and misleading images.
We can choose to stop putting money in the purses of those who keep the jaws of the Fat-Cat Sugar Trap oiled and poised ready to snatch us into their sticky, poisonous grasp.

We can begin to delete the Sugar Virus and re-write the text in our Guide Books to Life.
We can slowly disentangle the sticky threads woven into our Onesies and our collective Onesy. Sticky, deceptive threads of cultural acceptance when eating so much sugar makes us stupid, depressed, overweight and potentially extremely ill.

We can redefine sugar in our dictionaries, delete words like 'common commodity' and use words like 'precious' and 'occasional' instead.

We can cultivate gratitude and train our brains to reserve sugar for limited use, just as was stated in the original declaration by the 'Pancreas of All'.

When moments of weakness are prowling behind us like a shadow, think of the grim, manipulative, black hole of disease and unhappiness and the increased risk of obesity, diabetes, metabolic syndrome, carbolic poisoning, heart disease, poor brain function, bone disease, rotten teeth, cancer, Alzheimer's disease and the Hashtag 83% attached to our kids.

If you think differently about sugar, your liver, your pancreas, your intestines, your brain, your synapses, your heart, your hormones, your whole body, mind and spirit will happily thank you for it.

Think Sugar? Think Chemical.

CHAPTER FIVE

THINK HAPPY FOOD

Let food be thy medicine
And medicine be thy food.

Hippocrates

THINK HAPPY FOOD

As a swift antidote to the black hole, this chapter is a 'Cornucopia Cabaret' highlighting some of the foods that are particularly good for our brains and for eating our way to a happier life.

We can eat our way to a happier life. What we eat really does affect our health, our moods, our behaviour, our thinking and consequently our attitude to life and the amount of happiness we experience.

There are many foods that can help to reduce sugar cravings, help us quit the habit completely and gradually adjust our palate back to normal. Foods that make us feel healthier, foods that help to regenerate our cells and organs and foods that promote good brain function and clear thinking; the paving stones on which we can walk our way through a happy life.

As with most things, there are different schools of thought on the quitting sugar thing. 'Action on Sugar' recommends a gradual approach whereas others suggest the direct action of overnight cold turkey.

Whichever route you choose to take, it's good to bear in mind when setting off on your pilgrimage, that it takes about 21 days to form or break a habit. After this amount of time the neural pathways of the old way of thinking begin to wither away as our focus has been re-directed elsewhere. As we continue with the new behaviour these new neural pathways become more engrained and embedded in our thinking processes and the task of eating less or no sugar becomes progressively easier to adhere to. Before long, it's no longer a task, we don't even think about it.

I've been doing a mix and match approach to quitting sugar.

Chocolate and I had a civilised last supper. I booked it in my diary and prepared my speech; after all we had been lifelong friends. I expressed my thanks to it for always having been there for me but explained how I couldn't carry on with the deception anymore. I went on to say that I could no longer contribute to the devastation this cash crop wreaks upon our health and our Earth. Finally, I admitted that I had met someone else and that I'm in a new relationship with Blueberry.

Chocolate did try to catch my eye when I was in the supermarket, and I must admit I felt a little wrench as we had been such good friends, but I remained resolute and pretended that I just hadn't seen him. He got the huff in the end, and nowadays keeps out of my way completely when I go shopping.

As for sugar in my coffee, I've taken the gradual slope to happy valley and followed the recommendation from Action on Sugar:

'To cut consumption to half in the first instance and then gradually reduce intake from there'.

It took me a couple of months to go from two very generous teaspoons of fruit sugar (which is probably equivalent to at least three buckets of regular sugar) to absolutely none at all. I am now totally used to my coffee tasting disgusting without sugar. So much so in fact, that it no longer tastes disgusting, it tastes fine. In truth, now that my taste buds have adjusted, I actually prefer it and I get to do a little jig around the kitchen while singing a happy ditty proclaiming that, "There ain't no sugar in my Fair-Trade coffee".

There's plenty of good, sound advice available for reducing sugar or quitting completely. I do recommend the Action on Sugar website and I also recommend having a nosey about online; there are many encouraging stories, strategies and recipes to eat our way to a healthier and happier life.

In the meantime, welcome to:

THE CORNUCOPIA CABARET FOR HAPPY BRAINS
First up on the stage is the cheerful posse of:

THE COMPLEX CARBS
We don't normally think of our brains as being hungry, but hungry they are. Our brains are

constantly gobbling up glucose from our bloodstream and they use one fifth of all the blood that is pumped by our hearts.
The best way to keep the hungry brain happy is to have a slow and steady release of the glucose grub it likes best. Fresh vegetables, fresh fruit, whole grains and whole cereals all come naturally packed with a variety of complex carbohydrates that our bodies convert into glucose, which is then released slowly and steadily into our bloodstream, keeping the brain happy and functioning in a stable way. (1)
Refined flours, refined grains, processed cereals and processed foods have a quite a different effect on our blood sugar levels as they release a sudden burst of energy followed by absolutely none at all. This does not make for a happy brain. The brain gets irritable and grumpy, slopes off in a mood and refuses to concentrate properly until we feed it more grub.

Brains definitely work best if they've had their breakfast. No breakfast means a grumpy brain.

Next up, we have:

THE FRIENDLY EFAs
Essential Fatty Acids are vital for a healthy and happy brain. Brains really sulk if they don't get their EFAs.
EFAs are powerful and versatile nutrients that are necessary for transmitting signals between brain cells and have a significant impact on the quality of our brain function.
When a brain is weighed (not while still inside a head), about 60% of its 'dry weight' is made up of fat. A third of this is made from the EFAs omega 3 and omega 6, which our bodies do not manufacture so it's essential that we include them in our diet.
We have equal amounts of these omegas in our brains and many nutritionists recommend eating them in equal measure.
Omega 6 - eggs, poultry, avocados, nuts
Omega 3 – salmon, mackerel, herring, sardines, trout, pilchard, kippers, linseed, flaxseed, pumpkin seeds, walnuts.
(1)

Next up on the Cornucopia Stage is a world famous performer:

WONDERFUL WATER
Water is also essential for a happy brain. Brains are not only very hungry organs, they are extremely thirsty too. Our brain performs significantly better when it's had plenty to drink and is fully hydrated. Our brains (while still in our heads) are made of about 80% water. (Weird).

We use about 2.5 litres of water every day breathing, sweating and communing with the toilet, and we need to drink about 1.5 litres of fluid to remain happily hydrated. The rest of the water we need comes from the food we eat and from the chemical reactions that take place in the body. Contrary to popular opinion, the required 1.5 litres of fluid intake does not include booze.

Our brains are particularly happy when we give them the very stuff they're mostly made of, especially a couple of glasses in the morning as this kick-starts the brain in time for work, replenishing the brain and all our organs. Similarly, a glass of water before sleeping helps to keep us hydrated and prevents cramp during the night. (1)

Next up under the spotlight for brainy power with Brilliant Broccoli on lead vocals is the magnificent ensemble:

LEAFY GREENS & THE CRUCIFEROUS VEGGIES
Broccoli is one of the brain's favourite veggies as it contains two vitally important nutrients that assist our brains to function well, vitamin K and choline. These crucial nutrients significantly help to improve cognitive ability and memory, and given that us lot eat so much sugar we could probably all do with a daily dose of this power punch. Broccoli also contains a good dose of folic acid, which is important in helping to keep the brain happy as a lack of folic acid is linked to depression and Alzheimer's disease. (2)

Other members of this delightful ensemble are:
Cauliflower - (Violin)
Cabbage - (Cello)
Kale - (Tenor Sax)
Brussel Sprouts - (Not just for Christmas) and
Salads - (Percussion)

These fantastic vegetables are all power-packed with vitamins, antioxidants and carotenoids, and are especially good for a happy brain. We need antioxidants to prevent the damage caused by free radicals. Free radicals get a lot of deserved bad press because they are leftover, toxic, waste products generated by our cells as they convert fuel to energy (cell-poo). The brain is a hungry beast and even though it amounts to about 3% of our total body weight, it gobbles up 17% of our energy. Because the brain uses so much fuel, it also generates a lot of poo - the badass free radicals. Eating plenty of antioxidant rich vegetables helps to keep these bad boys at bay and our brains very happy. (3)

Please put your hands together and welcome our next act:

THE MERRY BERRY BAND
With Blueberry centre stage.

All berries and vegetables with deep colours help to keep our brains happy by promoting healthy communication between brain cells. They are also full of our friendly, antioxidant refuse collectors.
Blueberries are particularly power-packed for a happy brain as not only do they improve learning skills and memory, they can actually repair nerve damage that causes memory loss. So there is hope for our sugar-addled brains after all. Blueberries play a significant role in improving our ability to learn and our brain's ability to remember all the stuff it likes to learn.
Blueberries are fantastic to take with you if you decide to go down the reducing sugar trail; they're easy to snack on, help reduce sugar cravings, gently persuade our palate to re-adjust to the beauty of natural sweetness, and they help

our brains to learn and remember new habits like not eating sugar. (2)

Fruit and veg are best consumed without the added chemicals that many fertilizers and pesticides contain. Our brain has got enough on its plate chasing off the badass free radicals without having a serving of 40 or so added toxic chemicals to contend with as well. If it's not possible to get organic fruit and veg, it is possible to remove some of the toxins by soaking the fruit or veg in a solution of salt water or cider vinegar for ten or fifteen minutes. After rinsing, these solutions don't affect the taste of the food but do remove some of the nasty chemicals.

Next we have an eclectic mix who have just flown in all the way from the Middle East and Mainland Europe:

FATS & OILS, NUTS & SEEDS
Fats and Oils have had a bad reputation since the 1970's when they were incorrectly charged with being the main drivers of obesity and associated diseases. We've got it stuck in our heads that fats make us fat. Not so, sugar makes us fat. (And stupid.)
We need to re-think our thoughts on fats and oils because our bodies put them to good use in many ways. They provide a ready source of energy, fuelling all of our cells and making us feel full. They all contain vitamin E, an important vitamin and antioxidant, which help to prevent cognitive decline. Fats and oils help to balance the level of sugar in our blood and they provide a means to absorb and use the nutrients of the soluble vitamins A, D, E and K. Fats and oils also make an important contribution to the manufacture of our body tissues and the production of necessary chemicals and hormones. (4)

We go through life thinking that we're the same person every day, but in actual fact, our cells are constantly being replaced. Eating a good supply of these good fats and oils ensures our cells get replaced with an optimum health and happiness factor.

AMAZING AVOCADO
Avocados are often misunderstood and misrepresented. The unsaturated fatty acids they contain are particularly good at keeping our brain cells flexible and help to improve the strength of the brain's muscle. (No wonder the brain's hungry and thirsty all the time if it's down the gym working on its six-pack).
These mono-unsaturated fats are also excellent for good circulation and lowering blood pressure, both of which make a significant contribution to a happy brain, functioning at its best. (2)

EXTRA VIRGIN OLIVE OIL
Extra Virgin Olive Oil is an excellent source of fat in our diet, lowering blood pressure as well as reducing cholesterol levels. However, like butter, this oil breaks down at high temperatures which makes it harder for our bodies to digest, so best served cold or used for low temperature cooking.

COCONUT OIL
Coconut Oil and Ghee are both excellent for cooking at high temperatures because their vital properties don't deteriorate when heated.
Coconut oil has many nutritional benefits. The fatty acids help to boost our brain function, which is always a bonus. The fat in coconut oil helps to keep us full, reducing appetite while increasing our energy expenditure and burning off fat; coconut oil actually helps us to lose weight. This oil also contains Lauric Acid, which helps to fight

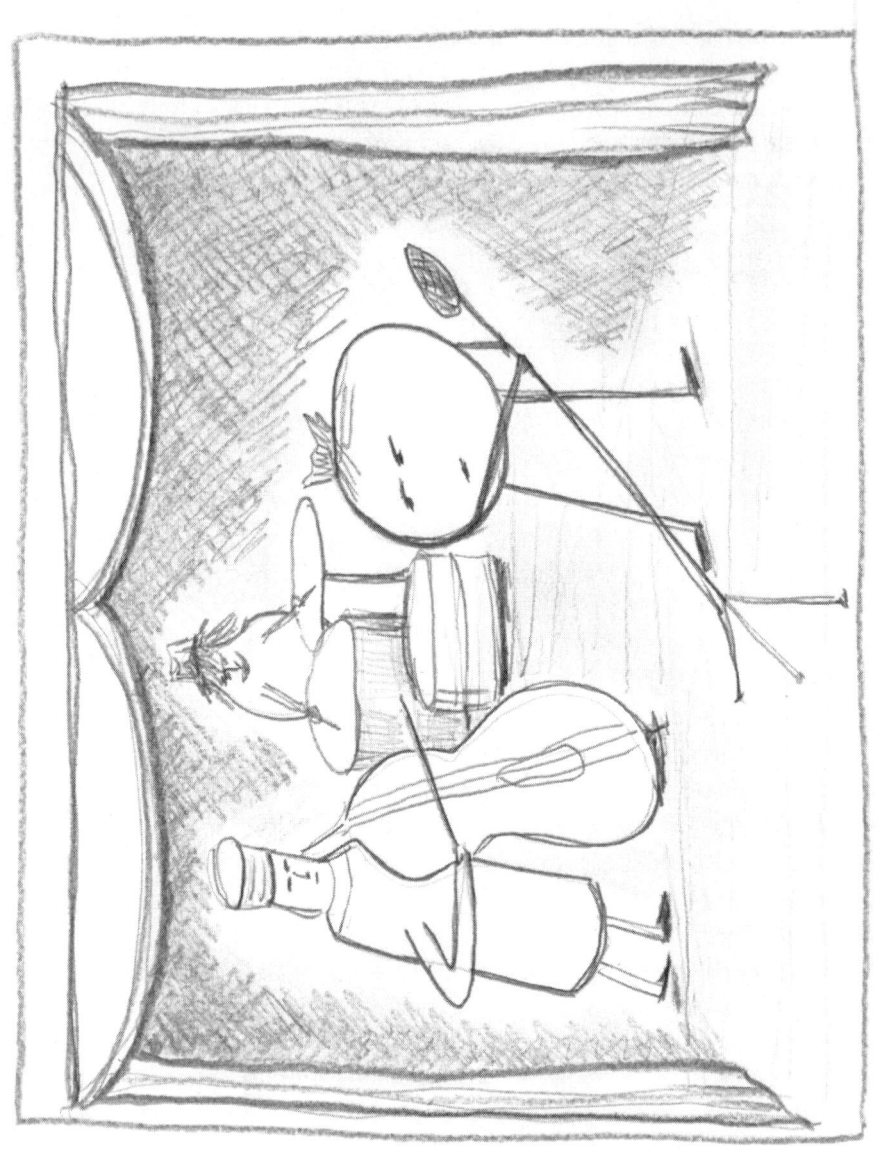

infection by killing off invading viruses or bacteria. Like Olive oil, Coconut oil also lowers levels of cholesterol in our blood, which reduces the risk of heart disease. (5)

The fats and oils to avoid are the trans-fats or hydrogenated fat or oil. Brains really don't like these types of fats as they prevent the friendly EFAs from doing their job properly. These sneaky dudes tend to hang out in cakes, biscuits, and other processed and manufactured products. Avoid at all costs. One good thing about my confrontation with the black hole is that I have finally picked up a cookery book and discovered how to make a cake. British life, it seems, is not complete without cake, preferably at 4pm. However, I've never been able to stick to a recipe, this time omitting sugar completely and using apples, dates and coconut oil instead. My Mum would be proud.
(If she wasn't dead.)

Now for a few solo acts.
The first is an all-time favourite, home grown vegetable:

THE TASTY TOMATO
The lycopene antioxidant content of tomatoes is especially good at protecting our brain from the badass free radical damage that can lead to dementia and Alzheimer's disease. Stick to fresh tomatoes whenever possible, eat raw or chop them, fry them, grill them, bake them, and scoff them. (6)

Next we have:

TARZAN TURMERIC
Turmeric contains a potent compound called Curcumin that has many nutritional and medicinal benefits which I

can't do justice to here. This powerful friend fights inflammation, is a strong antioxidant and contributes to good, happy brain function. The health benefits of eating Turmeric are wide-ranging: it's linked to a lower risk of heart and brain diseases and may play a preventative role against Alzheimer's, it's also linked to good brain function, combatting depression, it's a powerful ingredient in preventing cancer and can help with treating this disease. Turmeric may even assist with keeping us young at heart and help to fight age-related diseases. Check it out online. Turmeric is in many curry recipes and can be added to many other different dishes, dressings and marinades. (7)

Next up is a worldwide favourite:

COCOA
Cocoa has powerful antioxidant properties. But before you slope off to retrieve that secretly stashed bar of chocolate, double check that it contains at least 70% cocoa and, unfortunately, just one small piece a day is sufficient to gain the benefit of its antioxidant properties – no need to scoff the entire bar in one fell swoop.
Researchers in the field of nutrition have also found that cocoa improves blood flow to our brains, helping with verbal fluency and cognitive function in those of us who have had lots of birthdays. In this case, eating the entire bar is highly recommended. (1 & 2)

Next we have an exotic act all the way from the Far East:

GREEN TEA
This tea has a phenomenal reputation for being really good for our hearts, has preventative properties against cancer and more recently research has revealed that Green Tea is also good for our memory function and helps to keep our

brains concentrating and working at their best. This tea has an acquired taste, which I am still working on acquiring. (8)

Next to shine under the spotlight of brainy foods are:

THE AWESOME AMINO ACIDS
Amino acids play a critical role in our overall health. They form the building blocks of all proteins and make a significant contribution to our energy levels. They are also vitally important for our brain's neurotransmitters to do their job properly and keep our moods stable. Some of the amino acids are described as being essential because our bodies do not manufacture them and we have to be sure to include them in our diet. These essential amino acids are found in many different fruits, veg and plant-based foods as well as in meat, fish, eggs and dairy products. (9)

And now for the final act in our Cornucopia Cabaret, the magnificent community choir of:

VITAMINS AND MINERALS
Vitamins and Minerals play a crucial role in maintaining the good health of our whole body as well as that of our brains. Vitamins and minerals provide the necessary ingredients for our brains to convert amino acids into neurotransmitters and to transform carbohydrates into the glucose grub it likes best. A deficiency in vitamins and minerals leaves our brain under-functioning and not a happy bunny.
(1)
Vitamins also work in chemical conjunction with the enzymes in our cells to regulate the activity of our metabolism. Minerals like to get about a bit and contribute to a wide range of chemical and electrical activity that happens, unknown to us, throughout our whole bodies. (4)

Eating a varied and balanced diet ensures we get all of the vitamins and minerals we need. Many are particularly good for keeping our brains happy and a few are worth a mention here.

VITAMIN C
Vitamin C is well known for having the power to improve mental agility. This list shows mgs per 100g: (4)

Guava–230
Blackcurrants – 200
Spring Greens -180
Red Pepper - 140
Purple Broccoli – 110
Broccoli – 87
Strawberries – 77
Kiwi – 59
Lemon – 58
Mange-tout Peas, Clementines & Oranges - 54
Cabbage - 49
Nectarine & Mango - 37
Peaches - 31
ZINC
Zinc is essential for the brain to do its thinking and remembering.
Our brains have got a lot to think about and remember and by eating just a handful of pumpkin seeds a day, our brains will get all the zinc they need to keep thinking and remembering all the stuff they like to think about and remember, like eating pumpkin seeds, blueberries and not sugar. (6)

THE B VITAMINS
The B Vits, B6, B12 and folic acid are very important for keeping the brain happy and in good shape. These helpful

busy B's work to lower the levels of homocysterine in our blood, which decreases the risk of stroke, poor brain function and Alzheimer's disease. (6)

Good sources of B6 are:
Leeks, potatoes, avocados, red & green pepper, Brussels sprouts, bananas, cauliflower, broccoli, wheat germ, tempeh, muesli, yeast extract, wheat bran, tahini.

Good sources of B12 are:
Yeast extract, vegetable stock, soya milks, vegetable margarine, breakfast cereals.

Good sources of folic acid are:
Black-eyed beans, Swiss chard, savoy cabbage, spinach, broad beans, Brussels sprouts, lettuce.
However, please don't try to eat all of these lovely foods at once, unless you have a deal with the government to feed back methane into the national power supply. (4)

That wraps it up for our swift look at brainy foods, please put your hands together with a hearty show of thanks for all the wonderful grub on the Cornucopia Cabaret Stage for Happy Brains:

The Complex Carbs
The Friendly EFAs
Wonderful Water
The Broccoli Ensemble
The Merry Berry Band
The Amazing Avocado
Coconut and Extra Virgin Olive Oil
Nutty Nuts and Savoury Seeds
Tasty Tomato
Tarzan Turmeric

Cocoa
Green Tea
Awesome Amino Acids
Choir of Vitamins and Minerals

What we eat really does affect our health and our happiness.
If we eat well and keep our brains supplied with all the goodness it deserves, not only will we be much happier, our cravings for refined sugar will melt away into the past just as the morning mist melts in the heat of the sun.

Think Happy Food because happy food makes for a happy brain.
A happy brain makes happy thoughts.
Happy thoughts make happy feelings.
Happy feelings make happy actions.
Happy actions make a happy vibe.
A happy vibe makes a happy Onesy.
And a happy Onesy influences the Collective Happiness Factor bouncing about our planet.

Think Happy Food.

CHAPTER SIX

THINK SMILE

The arrival of a good clown in the village does more for its
health
Than twenty asses laden with drugs.

Thomas Sydenham - 17th Century Physician.

THINK SMILE

SMILING

'Women over fifty don't have babies because they would forget where they left them.'

As I'm a woman over fifty this joke made me laugh – so big thanks to whoever wrote it down (random find on the internet circa 2002).

Humour is a peculiar thing and particularly so because we don't all find the same things funny. Whatever your sense of humour may be, may it go with you because laughter is definitely the best medicine on Earth.

Just the act of smiling alone activates endorphin chemicals in our brains to do their happiness thing. A simple smile sends these chemicals whizzing about our body telling all our cells "Happy, Happy, Happy".

Smiling and laughter bring immediate happiness into our lives and a host of other wonderful benefits for our mind, body, spirit and Onesy: Our brain function improves with change of mood, relieving stress and anxiety, which in turn boosts our resilience.

Our physical health is bolstered because our immune system gets a boost of helpful hormones.

Stress hormones in our bloodstream are reduced and our muscles get a work-out stretching and relaxing when we laugh.

Pain management improves when we laugh and smile and this too helps to protect against heart disease.

Smiling and laughter also play a crucial role when relating to our family, friends, work colleagues and any new people we meet. Using smiling and laughter as part of our communication ensures we have good and lasting friendships because it's part of the essential glue that bonds us together. (1)

Smiling and laughter bond us together in a very powerful way. They create a positive energy that shines out from our Onesies and elicits a willingness of co-operation between people, between partners, in our families, with friends, at work and in our society.

UNIVERSAL LANGUAGE
It feels good to laugh and especially good to laugh with friends.
Even though there are many different types of humour and differences in what we find funny, smiling and laughter are a universal language. There are hundreds of different languages spoken around the world, of which we probably only know a couple, but we can all speak 'Smile and Laugh'. A world expert on facial expressions, called Paul Ekman, found that smiling means the same to us all, no matter which language we speak or our cultural beliefs. (2) This universal language has been with us since the beginnings of humanity, before we had speech like we have today. Smiling and laughter are part of the primordial mix that makes us human, that binds us together in groups and keeps us bonded together. Without it we probably wouldn't have survived this long. Smiling and laughter are a vital part of being a social species. We're not the only species to use smiling and laughter as a significant part of our social bonding communication; scientists in this field of study have found that rats, puppies and apes all use their versions of smiling and laughter to bond together. If a puppy lacks this essential survival skill and doesn't learn to 'laugh', its behaviour will be misunderstood by other dogs, perceived as aggressive and the puppy will get attacked. Rats are highly intelligent, ticklish and very playful. They have a range of vocal expressions, some of which indicate whether they are playing or fighting. Chimpanzees and

apes similarly use smiling and laughter to bond together and as a means of identifying friend from foe. (3)

It's probable that many mammals may have 'smiling and laughter' as a way of communicating; cats like to purr, for example. We humans are a very social species and smiling and laughter play an extremely important role in keeping us healthy, happy and connected right from the moment we were born. Modern ultrasound technology has revealed that babies appear to be smiling in the womb. Makes sense to me, all snug and cosy, no pressure to do anything except sleep and grow; reason enough to smile. Our brains are wired for smiling and laughter, even before we get born.

ENDORPHINS & CORTISOL

When we smile, the movement of our facial muscles kicks some of our brain cells into action that trigger the release of endorphins, the feel-good neurochemicals, to spread throughout our body. While these friendly endorphins are whizzing about our bodies making us feel good, the level of the stress hormone cortisol in our bloodstream is reduced. Cortisol, a vital hormone, has many important jobs and it naturally fluctuates throughout the day. The highest levels of cortisol occur in the morning to wake us up and get us out of bed, and lowest levels are at night while we sleep. Another crucial job that cortisol undertakes is to helps us to deal with stressful situations. When we're feeling anxious, worried or stressed, cortisol is released into our bloodstream from our adrenal glands. However, our bodies are not made to cope with constantly high levels of stress or high levels of cortisol. The pressure we are all under in the world today is causing many of us to be experiencing too much stress, too many worries and too much to do. Too much pressure and consequently too much cortisol whizzing about our bodies has serious adverse effects on our health.

High cortisol levels in our bloodstreams lower our immunity because it stops our white blood cells, our inbuilt defence system, from working properly. Too much cortisol can cause our bodies to deposit fat on our bellies, upper back and neck, and in order to lose this extra weight it's essential to address the underlying issue of stress.

High cortisol also causes a breakdown in our bones, muscles and connective tissue in order to increase blood sugar to the brain. It can also interfere with the thyroid system working properly.

Prolonged exposure to stress can lead to high cortisol spikes in the evening and during the night, which is one of the main causes of insomnia, waking several times in the night and night sweats. (4)

When cortisol is doing too much of its thing throughout our bodies, it triggers a set of associated negative thought patterns as we run through all the ifs, buts, whys and expletives of any stressful situation. These thoughts generate a pattern of negative feelings, and if we don't take hold of our thoughts we can easily get caught in a vicious circle of cortisol anxiety.

One way to take hold of our thoughts and catch this cycle before it starts spinning out of control is to smile. A simple act but actually quite hard to do when we're feeling totally stressed out. But it's worth a try because the response in the brain is directly linked to the movement of our facial muscles around the mouth and eyes and the brain will automatically release our mates the endorphins into our bloodstreams. We automatically get a blast of the feel-good factor simply because we made our face move, not because we saw or heard something funny or delightful. This part of our brain doesn't ask questions. If the face moves to smiling position, then endorphins get released.

End of. So even if we're in a grump, or especially when we're in a grump, if we smile our brains automatically activate endorphins to do their happiness thing despite feeling grumpy and we can hook into a different network of thought.

Back in the day, Charles Darwin was onto this way of thinking with his 'Facial Feedback Response Theory'. He recognised that the good feelings we experience when smiling are a response to the movement of our facial muscles, rather than us smiling because we feel good. (2)

PAIN MANAGEMENT

Smiling itself is a good enough reason to smile. The endorphins released into our bloodstream not only help to lower cortisol levels, they also act as the body's natural painkillers. Smiling and laughter have been proven to help with pain management and chronic illness. This connection between laughter and its effect on pain management and health became well known through Norman Cousins' best selling book 'Anatomy of an Illness' published in 1979. He was a man with remarkable tenacity as despite being given only a few months to live when diagnosed in 1964 with Ankylosing Spondylitis, a rare heart disease, he lived for another 26 years until he died in 1990. Norman Cousins' determination had him consuming large doses of vitamin C together with a massive injection of laughter and he kept on living. As a result of Cousins' book and remarkable recovery, considerably more research has been undertaken investigating the connections between smiling, laughing and good health. (5)

WELLBEING

The psychology department at the University of California undertook a research study that spanned some thirty years investigating the connection between smiling and wellbeing. Using the photos in an old yearbook, the researchers measured smiles and were able to predict the levels of wellbeing that the participants would experience throughout the course of the study. These predictions included: the level of fulfilment experienced in marriage and how long it would last, how each participant would score in standard tests of happiness and wellbeing, and to what extent each participant would inspire other people. They could determine all this just from a person's smile. Overall, this thirty-year investigation proved that those with the broadest smiles had happier lives, they had happier and longer lasting relationships and they lived longer. (2) Smiling is an easy way to acquire more birthdays and have a better time along the way.

2000 BARS OF CHOCOLATE
Quite remarkably, in a different study, some scientists have calculated that seeing the smile of a child is equivalent to having 2000 bars of chocolate. That's a lot of chocolate, even for the most hardcore of chocolate fans and midnight Choco-Walkers. 2000 small bars of chocolate at 36g per bar contains 8000 teaspoons of sugar. That's a very big pleasure fix just from seeing a smile.
This study was led by psychologist Dr David Lewis, author of 'The Secret Language of Success' and was backed by the well-known computer company, Hewlett Packard. In the clinical tests, participants were wired up to clever machines monitoring brain and heart activity and their responses to different stimuli were measured for 'mood-boosting' impact. The participants were given different stimuli: chocolate, cash, and photographs of their family and friends smiling. From this, the research team could measure different levels of stimulation and found that seeing the smile of a child

generated the same level of heart and brain activity as being given £16,000 in cash or eating 2000 chocolate bars. Feasting our eyes upon the smile of our beloved equates to about 600 bars of chocolate or a fat wedge of £8,500 in cash.

Seeing the smile of a friend scored £145 in cash or 200 bars of chocolate. (6)

HALO EFFECT

Exchanging smiles with significant people in our lives sets strong, positive emotions into action, which are accompanied by a string of chemical reactions in our brain, referred to as a 'halo' effect.

This is an all-encompassing effect, which taps into the positive networks in our brains and enables us to remember other happy moments more clearly, as well as generating more positive feelings of optimism and motivation. (6)

That's very cool; we get a Happy Halo just from smiling with our mates.

CONTAGEOUS

One of the quirky things about human nature is the way that yawning is contagious, and similarly, smiling has the same impact on our brains; it's infectious in a way that we have little or no control over. We've got an override programme embedded into the hard-drive of our subconscious mind that responds automatically when we see another person yawn or smile. We can't delete this programme or put it in the trash. It's our innate way of showing our friendship and mimicking another person's smile also allows us to understand something of what that other person is feeling.

Researchers at Uppsala University in Sweden (2) investigating this part of our hard-drive wiring, found that the control we normally have over our facial muscles is actually suppressed when we see another person smiling. We are compelled to smile. Not only are we demonstrating our allegiance, we are also getting a gauge on how the other person is genuinely feeling. By copying another person's smile in this way, we experience it physically in our own bodies and this gives us an indication of the other person's true emotional state.

The research in this field also demonstrated that it's very hard for us to frown when we're looking at someone who is smiling. (Teenagers have clearly been practising their control skills in this area and have developed an ingenious inner-app to override this particular feature of human nature that smiles when smiled upon).

FALSE SMILES

Our ability to mimic each other's smiles also serves as a means to detect a 'false' smile. Smiles that do not involve the muscles around the eyes are readily detected. Other false smiles may be less easy to detect at a glance and by mimicking each other we can gauge the other person's integrity. False smiles are not the same as genuine smiles and because we register that it's not a real smile, we associate these smiles with negative qualities, in particular a lack of trust.

Following on from the clinical research with chocolate, cash and photos, Dr Lewis and his team undertook another study where they showed members of the public photographs of the smiles of Politicians, Celebrities and Royals without disclosing any information. The participants were asked to give a score to each smile for different qualities, including warmth, honesty, sincerity and trustworthiness.

Interestingly, the smiles of Politicians scored the worst, especially for lacking in trustworthiness. (6)

RE-WIRES OUR BRAINS
The more we smile the easier it becomes to smile more often. The more we smile, the easier it becomes for our brains to make positive associations and patterns in our thinking. The more we smile, the more often our brains trigger the Happy Halo Effect, which fills us with optimism and motivation. While our brain is engaged with this way of thinking, it isn't thinking negative thoughts or pumping up the cortisol stress factor through our bloodstreams. If we increase the number times we smile we really do change the wiring in our brains. (7)

Some days I experiment with smiling at other people when I'm down the road at the shops. I really enjoy the way most people automatically smile right back at me. I find this a particularly helpful exercise when I'm in a grump. The act of smiling makes me feel better and seeing others smiling back makes me smile even more, putting me back on track and walking with happiness in my footsteps.

Real, beautiful, genuine smiles make our Onesies radiate happiness. Smiling makes us look good, approachable and competent in the eyes of others and this smiley–power goes all the way inside too, right down to a cellular level.

Smiling makes us feel happy, feel good, re-energised, more optimistic, more motivated and more creative. Exchanging smiles with another gives us a power boost and smiling to ourselves works just as well because our brains don't know the difference.

If you have no reason to smile, smile because smiling is one of the sweet elixirs for a long, happy and healthy life.

Children on average, smile 400 times a day. The happiest amongst us adults smile between 40 and 50 times a day, and the average amongst us smile 20 times a day. (8)
I reckon the average amongst us could probably smile a bit more.

If enough of us practice smiling every day, we could start a smiling revolution because smiling is more contagious than the common cold. We could at least have a National Smiling Day, not to raise money for charity, but to raise our National Happiness Level (NHL) raise our sense of connection and belonging, raise collective optimism, raise our national immune system (NIS), raise productivity and creativity, inspire each other and have more fun.
Smiling more often could in fact, save us humans from destroying ourselves.

LAUGHTER

Laughter is probably the most powerful elixir of all.
Although we're not thinking about any of the benefits when we're laughing, it's highly beneficial for us and the best free medicine for anyone. The good feelings stay with us long after the laughter has subsided and our aching belly muscles remain testimony to our bodies getting a work out. Along with stretched and aching abdominal muscles, our lungs heartily inhale gulps of fresh air replenishing all of our cells. This dual action stimulates a state of healthy equilibrium throughout our whole body, mind and spirit.

BRAIN WORK OUT
Laughter is good for our bodies, good for our souls and particularly good for our brains. It's an extremely quirky emotion triggering a response in our brains that travels in many directions simultaneously. Scientific research has found that laughter is the only emotion to do this. Generally

speaking, different areas of the brain are responsible for different tasks and the activity of emotional responses takes place in the frontal lobe of the brain. Laughter, however, is generated via an electrical circuit that travels along many pathways throughout several areas of the brain. Laughter gives our brains a work out as well as our bellies.

In the experiments led by Professor Peter Derks, College of William & Mary, USA, participants were connected to an EEG machine (electroencephalograph, which I can't pronounce either but with a big long name like that it's clearly a very clever machine), which measured the activity in the participants' brains when they laughed. The team discovered that our laughing brains all do the same thing.

In the nanosecond after hearing the end of a joke and before verbally laughing, our brains generate a consistent electrical pattern involving several different functions of the brain: laughter makes the left side of the cortex in our brain think about the actual words that are being said, activates the frontal lobe part of our brain that's connected to social interaction and emotional responses, kicks the brain cells into action on the right side that have to work out why the joke is funny so that we 'get it', the wave of electrical activity then extends to the part of the brain that processes sensory information, and then the motor section of the brain activates the physical response in our bodies: the funny noise we make and our bellies wobbling about uncontrollably. (9)

Laughter is the only emotion to fire all five cylinders at once and with good reason; it's designed to be extremely good for us. Laughter and a sense of humour make a significant contribution in keeping us healthy as well as happy, connected, motivated and inspired.

PEP UP OUR PEPTIDES

All of our emotions and thoughts are translated into a variety of Chemical Packages when stored in our brain and bodies. As our emotions change and fluctuate, the chemical information that is released throughout the body corresponds accordingly. Included in these packages are some very talkative fellows known to scientists as 'neuropeptides'. These complex molecules, which include the friendly endorphins, are made from amino acids and provide our cells with a way to talk to each other; a cell-phone for our cells. These neuropeptides are like teenagers constantly sending texts on three different networks simultaneously: brain-to-brain, brain-to-body and body-to-brain.

Cells in our brain, immune system and other cells throughout our bodies, have 'receptor sites' for these peptides to land on. Depending on our thoughts and emotions, different neuropeptides will whizz into action and go about their business saying different things, in response to our moods and thoughts.

Having a healthy dose of happy peptides whizzing around these receptor sites can actually affect whether we catch a dose of the latest bug that's going around, or not. The only way a virus can get into one of our cells is to land on the same receptor sites that the peptides use. If a receptor site is already occupied by its natural peptides, for example a shoal of endorphins, they act like bouncers trained in Martial Arts and the pesky virus doesn't stand a chance of getting in. (10)

If we keep pepping up our peptides with power-smiles and laughter, not only will we feel happier, we'll be right on course for staying healthier too. A sense of humour is a free supplement to include in our daily diet to ensure good health and to keep a flurry of happy chemical texts landing in our receptor site inboxes.

We can easily smile more often and with a smile on our face it's easier to welcome more humour into our lives and easier for our brains to access positive networks that make us feel optimistic and motivated.

EXPRESSION
The reality of life is that it presents many different experiences and emotions. We're capable of feeling an entire spectrum of emotion and it's important for good health to find constructive ways to express ourselves. Suppressed emotions get stuck in our bodies and have an adverse effect on our thoughts, moods and health. There's a simple truth in the expression, 'You'll feel better if you get it off your chest'.

It is much healthier for us to express how we are feeling, talk constructively about our grievances and find creative solutions, rather than bottling them up inside The Cupboard of Doom, to deal with at a later date. Unexpressed or unresolved negative feelings stuck in our minds, bodies and Onesies hold us back and prevent us from fully engaging with life. Storing emotions in The Cupboard of Doom creates a blockage that prevents a constant flow of happy through our bloodstreams. Because emotions are stored in our bodies as Chemical Packages every time we remember or recall a grievance the chemicals associated with that emotion are sent whizzing round our body.

Some of us in the UK are a bit reticent to speak our minds. We are known for our 'stiff upper lip' and not showing our true feelings.

While being polite and courteous are admirable qualities and are often accompanied by kindness, if we've got grievances rumbling about our blood and bones it's not doing our health any good. Whenever the grievance is grumbling so too are the grievance chemicals, upsetting the natural state of equilibrium that our brains, bodies and hearts like best.

If we let this unresolved stuff keep rumbling around in the Cupboard of Doom, eventually it could make us ill.

Writing about thoughts and feelings helps immensely. It can help to construct a potential conversation, it could be a letter or it can be a means to an end in itself; the act of writing serves to 'get it off your chest' and clears the stuff bottled up in your heads, hearts and peptides.

If you still don't feel better after that, I recommend that you savour your one piece of at least 70% cocoa chocolate, eat an entire punnet of blueberries, make a list of all the wonderful people in your life, smile your most beautiful smile, tell King Critical to feck off, laugh for taking yourself too seriously, put on your dancing shoes and turn the music up loud.

CONNECTION

Laughing with others is particularly good for us. We generate more laughter because things are funnier if we've shared the experience with someone else or a group of people.

Laughing with others builds positive, durable bonds and adds emotional strength to our relationships, injecting our friendships with vitality, playfulness, resilience and trust. Laughter and a sense of humour help us to deal with stressful situations and disappointments in life. It can help to gain a different perspective on disagreements and could diffuse a difficult situation in a jiffy. Laughing is cathartic and helps to repair any hurt feelings, disagreements or resentments that might be hanging around, bringing us back to a place of understanding and connection.

Laughter can shift our perspective enabling us to give

152

voice to awkward feelings, thoughts or situations between family, friends or colleagues.

Laughing together unites us at any time and serves as a powerful tool to unite us during difficult times, assisting our brains in staying alert, focused and on the case.

WORRY PACKAGES

Laughter fuels our brains with an energy that leaves a residual happiness effect that thinks via the positive and motivated networks of our minds. When our brains are thinking in this way it allows for more spontaneity, inspiration and creativity precisely because we aren't thinking about the stuff in the 'Worry Packages'. The Worry Packages include, and are directly connected to, the 'Package of Limiting Beliefs', the 'Package of Self Doubt', the 'Judgement Package of Criticism and Comparison', the 'Defensive Default Package', the 'Package of Procrastination' and the 'Package of Fear'.

With good humour coursing through our bloodstreams, our brain is working on a different network completely. The more we tap into this positive network the easier it becomes for our brains to get a good mood signal and have unlimited broadband access that can stream all manner of creative and happy possibilities.

One of the common barriers preventing us from inviting more happiness into our lives is when we get stuck in the Worry Packages. Life these days provides us with too much stuff to worry about and once we've opened up the Worry Package, we invariably find there's more worries lurking in the shadows at the bottom of the box, triggering connections to the other packages in this category.

Once the Worry Network has been activated, chemicals are whizzing around the entire body telling all of our cells "Worry, worry, worry", "Be scared, be scared, be scared". If we don't take hold of our thoughts, the associated chemicals and hormones will lock us into negative feedback loops.

Another link to the Worry Network is the 'Package of Self Importance'. When we take ourselves too seriously we can get stuck with a distorted perspective of our self, and from this viewpoint we can become very rigid in our thinking and rigid in our cells.
We're less likely to be able to think beyond the box and we're less likely to find creative solutions. We're more likely to miss opportunities for growth, learning, and an exchange of ideas, possibilities and happiness.

HAS AUNT MAUD DIED?
Often when we're worrying or react to something it's easy to project all manner of negative stuff into a situation or a future event.
In challenging, upsetting or stressful situations it's helpful to prod ourselves in the back and ask a few pertinent questions before we activate the Worry Packages and Networks of Fear.

Is the situation actually worth getting upset over?
Is it that bad?
Is it worth activating the worry chemicals?
Is it worth activating other peoples' worry chemicals?
Is it your problem or are you interfering?
Is it vital that something happens in a certain way or could it be done differently with just as much success? (Or even better?)
In the bigger scheme of things, how important is it?

Has Aunt Maud died?

TEN-MINUTE WORRY SLOT

An exercise I heard about a while ago, (apologies, source forgotten along with where I left the baby), is to get our entire day's worth of worrying over and done with in a swift ten minute slot. This leaves the rest of the day completely worry-free. If the Worry Package starts bleeping later on, we can easily remind ourselves that we've already done the Worry Slot for the day. Just as we can easily and effortlessly ignore unwelcome emails, we can do the same with our unwelcome worries. The more we practice not over-worrying, the easier it becomes and then we have less worries. No worries, mate.

GOOD HEALTH

Thomas Sydenham (he's our 17th Century physician who noted the sweetness factor in the urine of people with diabetes, and yep, he must have actually sampled some), also observed:

"The arrival of a good clown in the village does more for its health than twenty asses laden with drugs." (11)

He was an astute fellow indeed.

Our villages are bigger than in the 17th century and the asses have been replaced by 20 tonne lorries, but good humour wins the day whether four hundred years down the line or four thousand years, because laughter really is the best medicine for us all.

Smiling and laughter are part of our primordial mix that above all else, bonds us together, improving our relationships and quality of life.

Smiling and laughter ignite the Happy Chemical Packages, inviting endorphin hormones to whizz about telling our cells "Happy, happy, happy". Smiling and laughter relieve stress, anxiety and decrease the cortisol chemical packages telling our cells "Worry, worry", worry".

Smiling and laughter lower blood pressure, protect our hearts, boost the immune system and help with pain management.

Laughter improves circulation and exercises and relaxes our muscles.

Laughter fires up all five cylinders of brainpower and makes for a very happy brain. A happy brain creates a brain environment where all manner of inspiration and creative potential is possible.

Good humour, smiling and laughter are wonderful tools that inspire us with hope and help keep life and our sense of self-importance in a healthy perspective.

There is much in life that we cannot control, especially what other people do, what they say and how they behave.

We can control how we respond to other people and the very stuff that life presents to us. We can choose which network to activate, choose which server to subscribe to and pre-select our preferences in the Chemical Packages Menu.

Much of life is regular, same-old, ordinary, everyday stuff. In each moment of this everyday stuff we have the choice to think a bit differently. We have the choice to stick a smile on our face and give ourselves the chance to approach whatever we're doing with goodwill.

CRUCIAL

We are living in possibly the most challenging times since the Ice Age that wiped out the dinosaurs.

The difference this time is that it's our human behaviour that's causing a global crisis. To redress the balance we have to change our thinking in order to change our behaviour, and acknowledge our individual responsibilities in the choices we make.

The current global crisis is no laughing matter but our global community urgently needs a massive injection of happiness in order to engender more inspiration, connection, motivation and co-operation between us all.

So along with the crucial need to reduce our eco-footprint there's also a crucial need for us all to cheer the fuck up. Innit?

Think Smile ☺

CHAPTER SEVEN

THINK CREATIVE

Without Creativity coursing through our blood and bones
We are like a tree without roots:
We have no means to draw nourishment from life
And we're easy prey to the winds of pessimism,
Disillusion and resignation.

THINK CREATIVE

Contrary to popular belief, we are all wired to be creative. Just as we are born smiling, born to be happy, we are also born to be creative. Everyone is. We all have an innate ability to be creative and it's a streak that runs through all of life and all disciplines; the possibilities for creativity are unlimited.

BELIEF
The belief around our creative ability can get displaced if we've told ourselves that we're 'not the creative type' or if other people have said the same thing. The belief about our creativity can get knocked off kilter at an early age and can get stuck in our minds as if set in stone and a fact of life, just as the Sun goes down at night. Thinking that we're not creative places a limiting belief in our Personalised Guide Books to Life, and can stall the flow of happy in our lives.
We can re-write any limiting text because such a belief isn't set in stone, and nor does the Sun go down at night. We say the Sun goes down at night because that's what it looks like from where we're standing. The truth is the Sun doesn't go down, we're going up and round through the night skies while the Sun just carries on doing its thing at 483,000 miles per hour. (1)

If you have such a limiting belief and think that you're not creative, or if you think that the Sun does go down at night, please think again. We are all wired to be creative; we are all the 'creative type'.
If somewhere in your Chamber of Beliefs, there's a muttering from your internal dialogue saying you're not the creative type, that's exactly what's limiting you, the belief and the mutterings rather than your actual ability to be

160

creative. Challenge yourself to discover where and when you picked up the limiting belief and put it back down again. Challenge yourself to do something creative. Anything. Anything goes. Anything goes because being creative makes us happier, activating positive networks of thought and generating feelings that fill our body with life-affirming chemicals. While we're in this state of being we have better satisfaction with life and an improved quality to the time we spend at work, at home and with friends.

SOULMATE

Creativity is Happiness' soul mate.

Being creative makes us feel really good because we're a species who are designed to be extremely creative. Being creative brings purpose to our lives and fills us with a sense of achievement. Whatever it is, big or small, it makes us feel good about ourselves. Whether a creative approach to a mundane task or if we've created an actual something, we feel good with our achievement. We feel motivated and inspired to have a go at more of the same, or inspired to have a go at something else. Creativity puts our whole being in a happy, good mood and that happiness radiates into other areas of our lives, and significantly influences the Collective Happiness Factor by inspiring others to be creative too.

EXPRESSION

Being creative or doing everyday things more creatively allows us to explore different parts of ourselves and give expression to these parts of our identity. It gives us chance to think a bit differently about ourselves, think differently about our personal world and the world around us. If we're learning something new we have a sense of purpose, we gain competence and experience emotional happiness that accompanies creativity. If we're learning a new skill in

162

a group or workshop, we have a collective sense of purpose and gain friendships, which in turn can engender more inspiration and creative possibilities.

Giving voice to our Creativity Package re-sets our default mode, starts deleting limiting belief viruses and downloads the latest versions of: 'Purpose, Achievement & Competence'

'Inspiration & Motivation'

'Resilience & Happiness XX'.

In a study led by Paul Silvia, psychology department at University of North Carolina, students were phoned randomly throughout the study to report on their level of happiness. They found that the participants who came out tops were those who were doing something creative at the time of the phone call. The team's research also highlighted that we don't have to be a master or a maestro in order to harvest the positive emotional and motivational benefits. Even if our creative endeavours are a bit silly and impractical, or if they don't quite work, or even if they're completely off the wall, it matters not. What matters is that we have a go at something because it is the process of creativity itself that brings us so many nourishing rewards. (2)

FLOODGATES

Being creative ignites the engines of our souls.

It's a great feeling when we're in 'the zone' and whatever we're doing absorbs all of our attention. In this zone, things tend to fall easily and effortlessly into place. We don't notice the time and when we do look at the clock, it is usually with disbelief. Being creatively absorbed in what we are doing not only makes us feel good, it is especially good for the old grey matter wobbling about in between our ears. We have billions of brain cells in our brains, many of which are semi-dormant and have been hanging around for ages waiting for us to wake them up by doing something creative. Once a few of these sleepy cells have woken up and smelt the coffee (without any sugar in it), they will enthusiastically wake up loads of other dozing brain cells and fire them into action too.

Once we start to think more creatively and find small yet satisfying ways to be creative, the floodgates of our neural networks begin to open. It becomes easier to think a bit differently and easier to allow the river of creativity to flow through us, rather than flowing past us with some other fecker rowing the boat.

Brains love to think in new ways and once we start thinking along the creative pathways more often, our Creativity Packages will respond with surprising enthusiasm.

Researchers looking into the nature of ideas and creativity have found that the more ideas we generate, the more bad ideas we are likely to have. Having lots of bad ideas is no bad thing. It means we'll also generate lots of good ideas because our creative and curious brains aren't ever satisfied until they've hit on the right idea or the best solution. (3)

ATTITUDE
Creativity is a constantly flowing source of nourishment, which we can tap into anytime and apply to pretty much anything in life. Being creative can mean undertaking a specific project, thinking up a completely new idea, or it can mean doing the same everyday stuff a bit differently. Thinking creatively puts us in a frame of mind where we can see, hear and sniff out new ways of doing the same old things or doing something completely new. It's an attitude that can spark new ideas to life and inject vitality into any area of our everyday life. Approaching whatever task we're doing with a creative attitude motivates us to engage more fully in whatever it is we are doing, keeps us

firmly in the present moment and stops us slipping off into the past, the future, or the Worry Packages. Engaging more creatively in whatever we are doing gets our brains thinking via the creative pathways and networks in our minds. When we activate the Creativity Packages our bloodstreams are filled with vitality and Imagination is allowed to play with the freedom to be inspired and think up all manner of ideas and possibilities.

TEAM BRAIN

If you, like me, have been thinking that the brainpower behind creativity resides solely in the right hand side of the brain, please be willing to think again. Although the two hemispheres of our brains are recruited to be responsible for different tasks, with language activating more cells on the left and reasoning more cells on the right, the creative process kick-starts many areas of our brains happily into action. Latest scientific research investigating which bits of our brains are involved with creativity have revealed that many regions of the brain are activated: left and right bits; bits in the middle; bits at the front; and bits deep down in the hallowed grounds of the subconscious mind.

The entire cycle of the creative process from when we first start thinking about a new idea or new approach, to getting ready, to giving it a go, to giving it another go but a bit differently, right through to the finished article or result involves many different parts of our brains thinking and talking to each other. Depending on what we're endeavouring to create, different regions of our brains are being activated, and together they willingly work as a team to get the task completed. (4)

Research into what our brains are actually doing when we're being creative is a relatively new field of investigation. To make an inroad into this new field, Siyuan Liu, Allen Braun and colleagues from Maryland USA and Rex Jung, University of New Mexico, measured the brain activity of Jazz musicians and Freestyle Rappers while they performed known and improvised pieces of music. From analysing the differences in brain activity during the performance of known scores compared to improvisation, the researchers found that three specific areas of our brains are active during improvised creative thinking:

The Attention Control Network – these are some of the brain cells you are using right now to read this book, and we use this network when we have to concentrate on anything, for example: reading, listening, and doing difficult tasks.
The Imagination Network is where we disappear to when daydreaming (which can be particularly active while at work), and it's where we construct mental images in our minds and think about past or future events.
The Attentional Flexibility Network is the part of our brain that has eyes in the back of our heads. Its job is to be aware of what's happening around us and be aware of what's going on in our brains, and is responsible for switching between the Attention Control and the Imagination Networks.

This research study, which Jung describes as 'an approximation', found that the process of creativity generally requires for the Attention Control Network to chill out a bit so that the Imagination Network gets chance to be more active, get inspired and allow for new ideas to take shape. The creative process increases activity in both our Imagination and Attentional Flexibility Networks. (3)

We can all be creative and we don't need to be a professional Jazz musician or Freestyle Rapper either, we just need to flick a couple of switches in the networks of our minds, jump in a boat with both paddles and go with the creative flow.

LINKS
Happiness generates creativity and creativity generates happiness.
When we're in a creative mode, the link between creativity and positive thoughts, emotions and happiness are quite apparent; we feel good and our Onesies beam with joy. Interestingly, research has validated that happiness boosts the potential for more creative thought, as the likelihood of us having a creative breakthrough is much higher if we were happier the day before. Positive thoughts and emotions such as joy, gratitude, hope, curiosity, serenity, good self-esteem and happiness are all positively linked to creativity.
These networks tap into each other and create a synthesis with our Creativity Chemical Packages, just as negative thoughts and emotions are linked to each other and trigger the Worry Chemical Packages into action, all of which send Creativity running to the nearest daydreaming bus-stop a mile down the road.

WIRED TO KEEP LEARNING
The positive state of mind that accompanies creativity engenders a way of thinking that's highly curious and genuinely interested in what's going on in the immediate environment and the world at large. Brains love to figure out new solutions, love learning new information and skills because brains are wired to keep learning. When we feed the curious and creative part of our brains we are rewarded with dynamic and empowering feelings that add zest to life, a sense of satisfaction, competence and the inspiration to acquire more knowledge or learn new skills. This state of being bolsters our resilience to the demands of

life and enables us to embrace different perspectives on our own lives as well as the world around us.

Feeding and nourishing the curious and creative part of ourselves puts us in control of the boat, puts both paddles firmly in our hands, making us feel fully alive, motivated and happy. (5)

The feel-good factor and sense of achievement stay with us for a long time after we've been doing something creatively. This feel-good factor of the creative flow can then spill over into the rest of the day and the other things we do. If we purposefully cultivate our creative streak, good feelings stay with us influencing everything we do, influencing people we engage with and making for excellent weekends.

The more we Think Creative, the easier it becomes.

CREATIVE SUSTAINABILITY

We are naturally a creative species and with billions of brain cells at our individual disposal we have huge potential for thinking very creatively. Between us all we've got innumerable brain cells, potent with infinite creative possibility and with an estimated 7.3 billion people on this awesome planet, our collective potential is exciting.

It's exciting because at this moment in our human evolution we have the knowledge, the skills, the technology and the incentive to create a healthier, happier and 100% sustainable world.

Necessity is the mother of all invention and Creativity is its father. Right now, with things the way they are at home and abroad, us humans have an urgent necessity to think creatively and get inventing.

Us humans have come up with quite a few bad ideas that for the planet as a whole just aren't working. Now is the time to come up with some extremely good ideas. We are infinitely creative and we want humanity to survive. We want a happy sustainable world to live in right now as well as ensuring there's plenty of planet left for generations and generations to come.

BILLIONS ON THE CASE
Encouragingly, billions of us humans are on the case. Denmark was recently in the news for being an extremely happy country and generating 116% of their electrical demand by sustainable means. They've also set a target for the rest of the world by producing 42% of their electricity through wind power. Wind power in the UK is now cheaper than nuclear power and the cost will continue to drop. Costa Rica is currently running completely on renewable energy, and Bhutan has pledged its whole country to be 100% organic by 2020 and will be the first country in the world to achieve this shining status.
It's not possible to do justice to the comprehensive list of constructive and sustainable activity that's already happening in the world, as it would fill more than a whole book in its own right; it would fill volumes of positively happy books.

DISTRACTIONS
In an interview with Tony Wagner, Innovation Education Fellow at Harvard and author of 'Creating Innovators: The Making of Young People Who Will Change the World' (Ref 6), he talks about his approach to getting kids to think differently in order to generate ideas for happy living without costing the Earth. The article also discusses how we can all embrace more creativity in our lives, regardless of age. It's never too late to give something a go and it's never too late to learn something new.
The older we get however, the more distractions we have acquired along the way. It's these distractions, and limiting

beliefs, which stop us from tapping into our creativity more often. Being creative provides us with a sense of purpose as well as a chance to fire up our imagination and to express ourselves through whatever we're doing or making. If for whatever reason we haven't given enough voice to this part of our nature, there can be a sense that 'something's missing' from our lives. As we get older we are inclined to fill this gap with other stuff: an array of friends who might be just as stuck as us doing the same old in the same old way, material goods that might bring a fleeting moment of satisfaction, the telly, computers and other devices, alcohol and alternatives, and we also eat to distract ourselves, we eat our way through boredom and that unidentifiable gnawing sense of 'something's missing'.

We distract ourselves from the 'something's missing' feeling and we distract ourselves from the pressure of 21st century life.
When I was a little kid, on the whole, I didn't really like telly that much. I particularly didn't like the Sunday afternoon matinee films (of the 1960's and 70's) and would view my family with consternation for watching telly when they could have been outdoors or doing something constructive. To my utter amazement when I was about 11 years old, a new programme appeared on children's TV called: 'Why Don't You Just Turn Off Your Television Set and Go Out And Do Something Less Boring Instead?' It was as if they had read my mind. I was also astonished that the people who made TV programmes would actually make one telling us not to watch the TV. Despite the irony, I was impressed. Someone was on to it back then at the beginning of the 70's when we only had three channels on offer. Today we've got hundreds of channels on the TV, we've got the Internet, we've got computers, smart phones, tablets, social networks, gaming consoles and all manner of other devices

of distraction that we, as well as the kids, could switch off a bit more often.

Spending too much time on these devices takes our attention away from the present moment, distracts us from being fully aware and shrouds our creative nature.

Todays' technology is awesomely fantastic and is a key factor in awakening our collective perception to the interdependent nature of humanity and the Earth we live on; we are one human race on one beautiful planet in a vast, incomprehensible expanse of infinity.

But we're spending way too much of our time allowing ourselves to be distracted by all the stuff technology provides on our TVs, computers, smart phones and other devices.

BALANCE

Amongst all the information that's bouncing round the planet, satellite to satellite and into our homes, is a huge amount of negative news. We're experiencing or we're surrounded by extremely distressing conditions through displacement, tyranny, war, disease, droughts, floods, climate change, suicide bombers and random acts of insanity. It's overwhelming. It's overwhelmingly stressful to bear witness to, let alone to be living in such extreme conditions.

While we have an acute responsibility to engage in what's happening in our world and actively contribute to ensure a fair and sustainable world for all, being constantly bombarded with too much negative news serves only to feed the Fear Packages in our hearts and minds. Too much exposure to negative news can cause us to get stuck in a Loop of Fear and our cells' cell-phones' start bleeping, "Be scared, be very scared". With scared cells in our bodies, we are easily caught panicking in the sinking sands of fear and project more negative possibilities into our personal and collective future.

Because we're constantly being exposed to such devastating news every day without an equal exposure to positive news, my suspicion is there's an underlying belief running through many people's minds, a negative belief that we're all doomed; humanity has fucked up big time, we've lost the plot, we're consuming the Earth out of existence and it seems that it's not just the weather that's gone crazy.

But.

It's only recently that this nihilistic outlook has been shimmering like a mirage on our not too distant horizon and we can choose to perpetuate it or not. The human brain is wired to think and think again to find the best solutions, especially in times of adversity. It would be more productive to believe in the creative power of humanity and believe we can clean up the mess we've made. Believe we can improve communication with constructive co-operation and continue dismantling the Tower of Babel.

We could choose to switch off electrical devices of distraction a bit more often and stop subscribing to the ideas and notions in the Loop of Fear Networks that perpetuate war, death, poverty and disease as inevitable side-effects of life. They're not. We can switch to the Networks of Co-operation and activate our Creativity Packages for the sake of our individual happiness and for the sake of humanity's survival. Without creativity running through our blood and bones we are like a tree without roots; we have no means to draw nourishment from life and are blown away on the winds of pessimism, disillusion and resignation.

Creativity Pledge

For the sake of Humanity's Survival I pledge to Activate my Creativity Package.

I pledge to listen to what my Creativity Package is Begging me to do.

I pledge to do the Creative Thing Every day.*

Tamara Peirson

14 Feb 2018

* but not necessarily all left-handed.

WIRED TO RE-THINK

Many companies worldwide have changed their thinking and methods of manufacturing for a sustainable world, forcing other companies to follow suit. Building laws and regulations are changing fast; France has just passed a law to ensure all new buildings have solar panels or gardens on their rooftops. We can grow food differently and many farmers are already on the case with a worldwide surge in demand for natural, uncontaminated food.

We can communicate differently too; billions of us around the world value happiness as the most important quality in life and billions of us believe that peace between nations holds a greater promise of happiness than arguments and war.

Us humans have an infinitely creative potential to think differently, we've been constantly re-thinking and re-creating since primordial time when we first began to explore life beyond the treetops. We're wired to re-think and wired to be creative or us humans would have been extinct a long time ago.

We can train our brains to Think Creative more often.

A small, creative shift in our individual thinking, how we talk to ourselves and how we approach life will have a positive, creative impact on those around us, and the way we collectively move forward.

CREATIVE PROCESS

Creativity is a streak that flows through everything in life. Just like smiling, it's a streak we can cultivate and bring with us to whatever we do. We can smile at everyday stuff and we can be creative with everyday stuff. If we want to administer ourselves with an anti-dote to the stress and pressures of life and revive our brains, bodies and Onesies with the happiness of purpose, we could set ourselves a creative project.

James Young, author of 'A Technique for Producing Ideas' (7) identified that there are five critical stages in the creative process.
The five stages follow a distinct order in our minds and starts with us collecting thoughts, ideas and possibilities surrounding the idea or project we want to do. Often, we don't get any further than this. We hope that one day, inspiration will come knocking on our front door and we'll jump out of bed and suddenly be totally on the case. In the meantime we allow ourselves to get caught up by busy schedules and devices of distraction. When we do get a bit further with a creative idea, we start to collect articles or books, video clips, news items, interesting facts and figures, or the actual materials to make something with. We jot down our thoughts on random bits of paper or in a more orderly fashion in one of our electrical devices. We start to build an Assemblage of Stuff that forms the foundation of our creative project.

After this, our minds like to ruminate on all the information and possibilities that are held in the Assemblage of Stuff. Our brains like to ponder from various viewpoints and brains like to shake up the Assemblage of Stuff, just like shaking a Snow Globe or Christmas decoration that has glitter suspended in liquid.
After this, our brains like to put the thoughts and ideas of the creative idea on the back boiler, leaving it alone while the glitter settles down. We have to allow time for our chauffeur, the subconscious mind, to drive us to our destination. There's no escaping this part of the procedure and there's no knowing how long it will take before we

arrive. It could be as swift as one night's sleep before a 'creative breakthrough', the next stage in the creative journey. This is when, as if from nowhere, a thought lands in our heads and we know exactly what we want to do and how we want to do it. Our chauffeur has brought us to the last part of the creative process, where our project can culminate and come to life.

Young stresses the importance of remaining patient with our idea or creative project at this stage. He also advises the invitation of constructive criticism as when we share an idea with others they too become inspired and can suggest possibilities and potential that we might have overlooked.

CREATIVITY PACKAGE

We are all wired to be creative and we all need to be creative. When we're doing something that we really enjoy, our Creativity Chemical Packages bolster our whole mind, body and spirit. Creativity chemicals fuel us in such a way that our everyday work is more productive, we are able to sustain our efforts for longer and at the end of the day we're feeling satisfied and go home with happiness stuffed into the pockets of our Onesies.

To welcome more creativity into our lives, firstly we have to listen to ourselves. In order to do this we might need to unplug the phone, the computer, the telly and all other electrical devices of distraction, we might need to hide from the family, partner, or the kids, and we all need to turn down the Attention Control Network and turn up the Imagination Network so we can hear exactly what our Creativity Package is asking to do.

Begging us to do.

Does it want to make something?
Does it want to learn something new?
Does it want to do an evening class?
Does it want to go to college?
Does it want to play an instrument?
Does it want to dance?
Does it want to sing?
Does it want to make a cake? (Sugar-free)
Does it want to make a garden shed?
Does it want to write that book you've been mulling over for far too long?
Does it want to craft a canoe out of birch bark, fill a rucksack with provisions and take off on an adventure through the wilderness of unchartered waters?
Or does it want more coffee?

Creativity particularly likes its own space to think, work and play.
A designated room or area where we can close the door on other distractions, anchor ourselves, and generate the right vibe to entice our Creativity Package out to play and in to action. Using the same space serves as an anchor, making it easier for us to step straight into the Creativity Zone.

PLEDGE
Next, we need a plan of action. The Creativity Package wants to hear us pledge our intentions so it can happily get on with its job of
Think Creative, rather than being stuck in the waiting bay and continually deferred to "One day, maybe".
Pledges can be crafted in all manner of ways. We could write a contract, terms and conditions included, or make a formal declaration to ourselves in the mirror, or we could climb a mountain, put a stake in the ground and make a

proclamation to the whole world. It doesn't matter how we do it but if we state our intent clearly our Creativity Packages will take us by surprise.

Choose a start date and stick to it. Estimate how long it will take and select an achievable and realistic end date to aim for. At any point along the way we can re-adjust our targets. Our Creativity Packages will remain happy so long as we stay on the case, whatever pace we're going at. Identify any resources or skills required to make the project happen with achievable strategies to put these elements in place. One resource that we all need, regardless of our creative pursuit, is belief. If we believe in ourselves and in our ideas, our Creativity Package will do absolutely everything in its power to make our dreams and visions a reality.

NO WORRIES
Many creative successes are born from a process of 'trial and error'. We give things a go and if it doesn't quite work, we give it another go or we try it another way until we hit on the right formula.
Embracing 'mistakes' or 'failure' as an important, valid part of the creative process opens up the playground for Creativity and reduces any pressure we might be inflicting on that part of ourselves. Feeling completely safe and not afraid of mistakes, our Creativity Packages will invite unexpected possibilities in to play.
Creativity especially likes to play.

WIRED FOR CREATIVITY
We can re-think our Creative-Capability Hard-Drive Programmes and delete any limiting beliefs like "I'm not the creative type". We can insert the words Purpose, Achievement & Competence, Inspiration & Motivation,

Resilience and Happiness XX into our personal Identity Dictionaries.

We don't have to be a master or a maestro at something because it is the creative process itself is that fills our lives with these qualities.

We could switch off electrical and other devices of distraction more often and make time to listen to what our Creativity Package is begging us to do.

I saw a quote attributed to the Dalai Llama recently, which is in fact taken from David Orr's book 'Ecological literacy: Educating our Children for a Sustainable World', saying that the planet doesn't need any more successful people because world societies have woven money, greed and exploitation into the definition of success.

$(H = LWH + M)$.

What our world needs more of, along with the peacemakers, storytellers, healers, dreamers and people of courage that David Orr suggests, is more of us lot being creative every day and going home with happiness stuffed into the pockets of our Onesies.

Our creative potential is unlimited.
The list of possibilities is endless.
Just pick one.

Think Creative.

CHAPTER EIGHT

THINK NOW

You should sit in meditation for twenty minutes a day,
Unless you are too busy,
Then you should sit for an hour.

Old Zen Proverb

THINK NOW

TIME

Time is a very peculiar thing and appears to speed up the older we get. When we were five, a year seemed a very long time indeed. At this age a year is one fifth of our existence, so relatively speaking it's extremely long and time does seem to stretch indefinitely.

It's probable that this erroneous belief that we have loads of time gets embedded into our subconscious hard-drive, along with all other the thoughts, feelings, sensory experiences and information we are constantly downloading and installing up to the age of seven. We grow up believing that time is extremely long and we can go through much of our lives thinking we have loads of time to do all the stuff we want to do, time to manifest our creative ideas, time to go to all the places we want to visit, time to change into the person we really want to be. Because we think we have endless time, often much of what we really want to do or be gets deferred to an indefinite point and we put it off until "one day".

By the time we get to fifty, one year is a fiftieth of our existence and time appears to fly by as if some sort of trickery is taking place, as if someone's been messing with the clocks and the calendars. In our adult lives we have increasingly more demands on our time. Often we're too busy to make time for the stuff we really want to do, or the person we really want to be. We have more commitments to meet, bills to pay, work to do, house, garden and family maintenance, and time becomes compact.

Time becomes squashed and brains become very full. Our brains are full of all the thoughts and feelings associated with the million and one things we have to do, places to be and people to see, crammed in together with all the thoughts and feelings of our dreams and visions, hopes and desires, worries, anxieties and fear. In our adult lives there's so much to squash into our time that often we become consumed by having to get things done. Our brains are so full that often we're doing one activity while thinking about the next task and the task after that. We're no longer present in the moment because we're thinking about all the other stuff we have to do.

One way we perceive time is as a linear progression from birth to death and all the stages in between. If I were to draw a line on the floor and ask you to stand on it and indicate which direction was past and which was future, you would more than likely point behind you for the past and in front of you for the future. (Not all people perceive time in this way, some may put the past in front and future behind, and some may perceive time as running laterally from left to right or vice versa). The 'timeline', whichever orientation, eventually meets up because life is a cycle. We end up where we started from; we end up somewhere other than here and my guess is that we are either nothing or we are just spirit, no body attached.

CYCLE OF THE PLANETS
Our collective perception of time is determined by the cyclical nature of the planets: the Earth spinning on its axis every day, the Moon orbiting the Earth every month and the Earth orbiting the Sun every year. The Earth spinning on its axis provides the gift of night and day, albeit in varying quantities depending on the time of year and relative position to the Earths' equator. The Moon's cycle is 27.3 days, which makes February the closest to a real moonth that we have. Many moons ago the year was divided into 13 months on account of the fact that 13 cycles of 28 equals 364 days. This is much closer to the actual movement of our night-clock than the seemingly arbitrary

thirty or thirty-one days of our current calendar. I blame the Romans. If they really wanted to do us a favour when they changed the calendar, they should have made January about ten days long or scrapped it altogether.

So here we are on Earth, which is rotating on its axis at a varying speed between 0 and 1,040 mph depending on where we're standing in relation to the equator. Here in the UK we're spinning at about 600 mph. (1)
At the same time as spinning on our axis, us lot on Earth are orbiting the Sun on our annual journey of 600 million miles, at a speed of 66,000 mph or 107,000 km/hr.
While we are spinning on our axis and hurtling around the Sun, the Sun too is blazing its way through our galaxy, the Milky Way. One lap of the Milky Way takes approximately 225 million years to complete (I timed it once a wee while ago), with the Sun moving at an astounding 483,000 mph or 792,000 km/hr. And the Earth, with us lot on it, is being pulled in the Sun's wake at the same phenomenal speed, albeit at a distance of 93 million miles.
Please forgive me if I'm wrong, but it seems that the Earth is moving at three different speeds simultaneously and in three directions simultaneously too. I'm sure physics has a mathematical equation involving a string of exciting equations to determine lots of interesting facts in which the end result probably includes good old gravity and somehow creates the illusion that we aren't moving anywhere at all. We live with the constant illusion that the Earth, with us lot on it, isn't spinning at 600 mph in the UK, not moving round the Sun at 66,000 mph and definitely not trailing behind the Sun at 483,00 mph.
But we are.
How physicists and mathematicians even figure out the sums for this sort of stuff, let alone do the maths and get the answers right, is beyond me. The same clever peeps have

calculated that since the formation of the Sun and our Earth, a time span of approximately 20 galactic years, the planets have done about 20 laps around our galaxy. From a different perspective, since the beginning of all recorded human history, we've barely progressed in the long journey through the Milky Way. (2)

Given that it took a wee while to set the stage and to get the conditions on Earth just right for an astoundingly diverse spectrum of life to flourish - 20 galactic years at 225 million each makes a total of 4,500,000,000 years - it seems more than a bit stupid if we blow it all before we've even got to 'Star Gate 1' or whatever we might call the various stages of the Earth's adventure around the galaxy.

You may have come across this following snippet of thought online:
The Earth is 4.6 billion years old; scaled down to 46 years, we have been here for 4 hours.
Our industrial revolution began 1 minute ago.
In that time we have destroyed more than 50% of the world's rainforests.

On the same timescale, in the next few seconds we could easily replant the forests and clean the rivers, the seas and the air, and replenish the earth.

ORIGINAL CREATION
Earth is a most beautiful, all-singing, all-dancing, self-sustaining, incredible creation. Whether it's been created through a random collision of elements that got battered into shape and life by meteorite activity, or whether it's been created by a divine being of your naming, or whether it's been created by consciousness itself seeking to

experience life, or whether our origins are a combination of all of the above, doesn't really matter.

What matters more than knowing the origins of life is that we get our act together and respect it. What matters most is sustainable wellbeing for this planet Earth and all of its inhabitants.

TOO BUSY

Meanwhile, leaving the origins of the universe in the domain of the unknowable for a while, here we are now and, 'Time stands still for no man, nor woman, nor for any of us who are a bit of both'. We're slaves to the clocks and chained to our work schedules. The list of pressures and demands of life in the 21st century seems as infinitely long as a Galactic Year. We're so consumed by our busy lives that we don't have time to remember how phenomenal it is to be alive. It's as if we've forgotten how amazingly lucky we are, because we're too busy and pre-occupied to notice. Busy with commitments, work, bills, shopping, chores, health issues, work issues, relationship issues, fitting in a social life, and then we're busy with devices and vices of distraction because it's all too much.

We're too busy to notice now.

Which is a really big shame, because now is where life exists. Now is where happiness lives.

REAL TIME

Yesterday and before are memories encoded in the neural networks of our minds. Getting out of bed this morning, a couple of hours ago or five minutes ago, is a memory. Tomorrow, next week, next month have yet to unfold. They are a vision of the future In the dreaming of our minds where we may place our hopes and desires, and where we project our fears and anxieties.

The past and the future only exist as memory or projected thoughts and real time is happening now.
But how much of our time do we spend somewhere other than in real time? How often do we spend time thinking about past events, what
we could have done, stuff we could have said, or thinking about what other people have said or done?

How much time do we spend thinking about future events, thinking about what we've got to do, and get, and be?
How much of our time do we spend being here and now?

Remembering people, events and places from the past seem real enough. When we recall memories the brain is busy reconstructing the images, the words, the feelings, the sounds, tastes and smells associated with the memory.
While our brains are busy re-constructing all this stuff, the associated Chemical Packages are activated to do their thing, although to a lesser degree than in the initial experience.
If we are remembering a moment of friendship or laughter then our mates the endorphins spring into action, fuelling us with the feel-good factor. Similarly, if we are recalling an unhappy or traumatic memory, the negative networks are activated and our bodies are filled with transmitters telling our cells, "Upset", "Stress", "Worry" or "Scared".
While our brain is busy doing all this reconstruction stuff and electrical and bio-chemical activity, and while we're busy looking at it all with our mind's eye and feeling emotions in our bodies, we're not experiencing real time in the here and now. We're experiencing the other thing. The same activity happens in our brains and bodies when we project ourselves into the future.
While daydreaming and visioning our future is as necessary as gleaning knowledge and understanding from our past,

in the 21ST C many of us are pre-occupied with too much to think about.

We can be like a vinyl record stuck in a groove playing the same thoughts over and over again, ruminating over the past or caught in endless worries about our future. And quite often when we're not stuck in those places, we're stuck in our smart phones, our computers, our TV's and other vices of distraction.

Many of us are missing too much of now.

Now is where life truly exists and where we can find Happiness happily playing in the sunshine, even if it's raining.

MEDITATION
Being present in the real time of now is exceptionally good for us. There are many disciplines, sports and activities that demand for us to be entirely focused in the present moment. Incorporating these types of activities into our everyday lives increases our mental ability to remain in the present moment and brings many benefits that boost our brains, bodies and wellbeing.

Meditation, an ancient practice found in Buddhism and Eastern religious traditions, is one such discipline that in recent times has migrated to the West. The non-religious, secular technique 'Mindfulness Meditation', which Jon Kabbat-Zinn adapted in the late 1970's from a Buddhist tradition, has become very popular in the UK, the USA and many other countries worldwide. This secular approach has paved the way for making meditation accessible for Western Monkey Minds, and because it works and is so very good for us, it's catching on big time.

In Buddhist and traditional meditation practices, placing focus on the breath serves both as an anchor and a resting place from which it is possible to go beyond thought. It's possible to enter the silence and experience an awareness that is beyond the confines of the physical body. In the realm and comfort of silence, there is a place of paradoxical awareness of existence and non-existence. The illusion of oneself becomes apparent because consciousness knows there are no boundaries. It's a place where consciousness perceives all and nothing. We are nowhere and everywhere at the same time, just as quantum physics describes.

Mindfulness meditation also uses focus on the breath as an anchor, to which we can return when our Monkey Minds swing from tree to tree and jump from thought to thought. An underlying principle of both Buddhist Mindfulness and Mindfulness Meditation is to just observe our thoughts without judgement. Notice the thought, notice the response and without giving ourselves any grief that we've been thinking something or talking internally, gently bring awareness back to focus on the breath.

A useful way to think of this is described in an article where mindfulness meditation was presented to school children, who were invited to imagine themselves waiting at a bus stop and each thought envisaged as a 'thought bus'. They could choose to get on the bus and allow their minds to travel with that thought or they could just let the bus pass by without hopping on board. (3)
Using visual images with a meditation practice can be very helpful and thoughts can be represented by anything we like – clouds, birds, tractors, ships, anything goes if it works as an anchor to maintain a meditative state.
There are many different approaches to meditation, different breathing techniques and there's a whole host of audio recordings and apps available to suit everyone. Try

some out, find something that flies your kite because meditation really is everything it's cracked up to be.

If you're thinking meditation is difficult and that you can't do it, or it's uncomfortable and you have to sit like a statue and not think anything at all, please suspend those thoughts for a moment while I fill you in with some of the reasons why meditation is exceptionally good for us, and why meditation is a path that leads directly to happiness.

ALL ROUND BENEFITS
Practising Meditation trains our brains to keep both hands on the steering wheel through life, trains our brains to acknowledge there are no back seat drivers and helps us access the source of happiness from within.
Practicing meditation is exceptionally good at tempering the voice of King Critical and guiding us to choose helpful options from the Think Think Menu. It helps us recover from The Pointy Finger Syndrome and is a fantastic tool for spring-cleaning The Cupboard of Doom.
Meditation is good for our all round health and happiness. It relaxes us, lowers stress and improves our ability to remain calm in all situations.
It improves our mental abilities of attention, concentration, learning and memory.
Meditation helps to preserve our aging brains and can help with depression and addiction.
Meditation improves awareness, increases our ability in relating well to others, improves relationships and our ability to empathise and be compassionate.
Meditation improves our overall sense of wellbeing, quality of life and happiness.
Because meditation does all this for us, it makes our Onesies cast a shining light into our grey and troubled world.

CHANGES OUR BRAINS

Recent scientific studies have confirmed that meditation does indeed bring us all these benefits. Science today has been able to confirm what practitioners of meditation have been saying for donkeys' yonks: about two and a half thousand years. (Which equates to seven and a half donkeys).

Using EEG and fMRI machines, scientists researching in this field have also been able to demonstrate that meditation physically changes the structure of our brains. New neural pathways and connections form as we train our brains to Think Now and relax into the meditative state. Meditation also strengthens the existing helpful and positive networks of our brains while weakening the unhelpful thought pathways of the Worry and Stress Networks.

Practising meditation regularly really does re-wire our brains. It does wonders for our overall brain health as well as increasing the volume of grey matter in specific parts of our brains.

RELAXATION

One of the biggest attractions of meditation for us busy and stressed- out Westerners is the feeling of relaxation that accompanies the practice. The relaxation benefits are immediate and long lasting. The more we do it the easier it becomes. The more we Think Now, the easier it becomes for our brains to chill out and swiftly enter a zone of deep relaxation. When we succumb to this state of being, our brains take a stroll along the Relaxation Networks in our minds, activating the Relaxation Packages to go about their business telling our cells "Relax", "Chill out", "No worries", "Think Now". While we are operating on this network, our Stress Packages are lulled to sleep and the level of cortisol in our bloodstreams is significantly reduced, bringing a host of benefits.

Because we're so busy and so stressed, many of us have got too many cortisol molecules whizzing about on their cell-phones saying "Schedules", "Deadlines", "Bills to pay", "Money to find", "Worry", "Upset", "Anger", "Tired", "Very tired", "Can't cope".

This stress, along with all the associated negative feelings, makes our bodies store fat and crave the very foods that make us put on weight. Meditation is particularly helpful for those who are suffering with stress in this way. (4)

SELF-REFERENCING MONKEY MIND

One of the ways meditation changes the internal networks of our minds is by reducing the activity in the 'self-referencing' part of the brain. This part of the brain is also known as the Medial PreFrontal Cortex by all the brainy people who research all this brainy stuff (5) and it resides within the Default Mode Network, DMN, which is an area of the brain more commonly known as Monkey Mind. Often, our self-referencing Monkey Mind doesn't know when to shut up. It's continually wittering on about everything we encounter that we deem as relevant to ourselves or to our lives in some way. This is where King Critical likes to hang out and where we can find The Cupboard of Doom.

We are constantly filtering and running information through the self-referencing part of our brains in order to make sense of the world in relation to ourselves. This part of our brain is on the case when we're talking to others, gauging how others might be feeling and responding accordingly, and when we feel sympathy and empathy. It's on the case when we're reflecting on ourselves, thinking about personal health issues, relationships, or any situation. It's busy at work when we're thinking over what we've said, running conversations back through our minds, reflecting on what other people have said and what they might think of us. It's also on the case when we're daydreaming or when we're thinking about future events and possibilities.

Our self-referencing centre of the brain has two parts: the first is called the Ventromedial PreFrontal Cortex (vmPFC), which is a bit of a mouthful, so I'm going to call it 'Maybe Mayhem'.

The Maybe Mayhem part of our brain is responsible for filtering all the stuff relating to ourselves, and all the people we perceive as being similar to ourselves. This is the part of our Monkey Mind that can get the wrong end of the banana. It's the part that goes Kangaroo Jumping to wrong conclusions. It's the part of our brain that can take things too personally and can send us off on a tangent thinking far too much. Thinking too much activates the Worry Packages or the Anxiety Packages into causing mayhem and leaving us out of balance with more cortisol molecules than happy endorphins swimming about in our bloodstreams.

The second part of our self-referencing centre is called the Dorsomedial PreFrontal Cortex (dmPFC), which I've changed slightly to: 'Mostly Manners'.

The Mostly Manners part of our brain is responsible for filtering all the information relating to people we perceive as being different to us. Mostly Manners undertakes the important jobs of maintaining social connections and for us to be empathic, particularly with people who we perceive as being different to us.

Practicing meditation quietens down our Monkey Minds, turns down the volume in our Maybe Mayhem sectors and simultaneously weakens the direct pathways from this part of the brain to the Amygdala, where the Package of Fear

Monkey Mind

I'm so worried, I can't sleep

Yeah. Me too.

It's terrible

Yep. Really awful

I'll activate the
Worry Packages

OK

I'm worried about so much...

I think I'm scared too...

Yeah. And me

I'm gonna activate the
Package of Fear now.

OK. Good idea.

I'm shit scared.

What was that noise?

Wtf? Is it a burglar?

OMG

It must be

GO AND LOOK!!!

No fnw

You go

fotnfwigdstlfabffs

No fnw

OK. I'll go & check

Phone the Police

It was the cat

lives, and weakens the unhelpful pathways to the Insula, where our Bodily Sensations Packages reside.

Our Package of Fear is very useful. Its job is to sound the alarm if we are in danger, activating our 'flight or fight' response, and often provides our first emotional response or reaction to something.

It's probable that the original programme for our Package of Fear was designed for infrequent use. The preferences originally stored were for special occasions only, such as an encounter with a bear, or a wolf, or a rampaging and rather unfriendly tribe. We're still running the same programme for fear and although in the UK we no longer have bears, wolves or rampaging tribes, our Packages of Fear are overloaded with far too much work to do in these stress-filled and demanding times. This part of our brains would particularly like to learn how to meditate, and to revert to its original factory settings and be reserved for occasional use only.

Our Bodily Sensations Package is running a programme that checks and double checks the sensations and 'gut-feelings' we experience in our bodies. Working with other bits of our brains, it will discern how strongly we respond as we ask ourselves if the sensation is something for concern or not. Another important role for this part of our brain is to activate our Empathy and Compassion Packages, where we feel and experience empathy and compassion in our whole bodies.

TALKING TOO MUCH

Owing to the nature of our stress-filled 21st century lives, the Maybe Mayhem and the Package of Fear parts of our brains are In conversation on a phone line far too often. They can make unlimited free calls and have the capacity to keep going on and on, all day long sometimes. They love repeating stuff over and over again, worrying about stuff they've done, blahing on about stuff they should have done, making comparisons and judgements of themselves and others, panicking about aches and pains in their bodies, and more besides. While we're caught in the hamster wheel of Maybe Mayhem, Maybe Worse, we are seriously lacking in Happy and we're completely missing the moment of now.

BRAIN BRAKES
Fortunately, we've got some brakes embedded into our brains.
Brain Brakes are usually referred to as the Assessment Centre or the Lateral PreFrontal Cortex.
Brain Brakes are the part of our brains that knows how to take a deep breath and count to ten. Brain Brakes know why it's important not to take things personally and knows when to take a walk up the mountain. It's the part of the brain that engages a more rational viewpoint. Our Brain Brakes enable us to evaluate from a position of balanced logic while tempering emotional responses that emanate from our Maybe Mayhem and our Package of Fear sectors. Our Brain Brakes have the power to override automatic responses and habitual behaviours while increasing our awareness in such a way that we don't take things personally.

Practising meditation on a regular basis cleans the rust off our Brain Brakes, installs new discs and pads, and boosts our stopping power. Practicing meditation weakens the phone line connections between the Maybe Mayhem, the Package of Fear and the Bodily Sensation Package sectors in our brains. Because these neural pathways are weakened when we meditate, our anxiety decreases significantly; King Critical is lulled to sleep and is replaced with a calm and centred perspective on our inner world as well as the world around us.

EMPATHY & COMPASSION

Another significant change that takes place in the meditating brain is that the parts involved with empathy and compassion, particularly empathy for people who are not like us, become much stronger. Our Mostly Manners sector gets much better at being able to work out other people's emotional state. We are more able to identify other peoples' reasoning, their needs and motives, their wishes and aspirations. With our Mostly Manners programme running in this way, we are able to discreetly slip into another person's moccasins and, with empathy surging through our bloodstream, we are able to express our true, compassionate nature.

15-MINUTE BRAIN BOOST

We really do re-wire our brains when we meditate; it's free and accessible to all with no soldering skills required. The more we meditate the easier it becomes. The beneficial pathways and networks get stronger and more powerful every time we give it a go. Fortunately, it doesn't take much to set these beneficial pathways into place and keep them activated. Meditating for just fifteen minutes a day is enough to maintain these Highways to Happiness, and it's enough to continue eroding the connections and conversation between Maybe Mayhem and the Package of Fear.

Brains love learning new things and brains particularly like to learn and practice meditation. For brains, meditation is akin to an all-expenses- paid, exotic holiday of your choice. Just as with any new task we give our brains to learn, the more we do it the easier it becomes.

(Yes, I am hypnotising you with my repetition).

The more we Think Now, the quicker our brains will remember what we're trying to do; they'll obligingly put on their Flip-Flops of Happiness and willingly stroll off into the sunshine of Happy Valley to top-up the suntan on their Onesies.

The pathways in our minds have a malleable quality that responds directly to our bidding. We just have to stay on the case or our brains will just go back to the old way of thinking. It's a bit like playing snakes and ladders; if we don't maintain our practice and keep the Highways to Happiness operative, we slide back down the snake and have to start from the beginning again.

COGNITIVE SKILLS, INNIT

Another magnificent benefit of the powers of meditation is the way it improves our cognitive skills of learning, memory, attention and concentration. A study in America (6), found there were significant benefits even after only a couple of weeks of meditation training. Students who undertook the training scored at least 16% higher in school examinations than those who hadn't practiced meditation.

Meditation improves our cognitive skills whatever age we happen to be. It is a state of being that readily transposes to many situations in life that demand our attention and concentration. Because the noise of our internal wittering from the Maybe Mayhem and Package of Fear sectors has been turned down or off completely, our awareness increases and our ability to focus in the moment of now vastly improves. Our focus in the present moment generates an environment in the brain that makes our

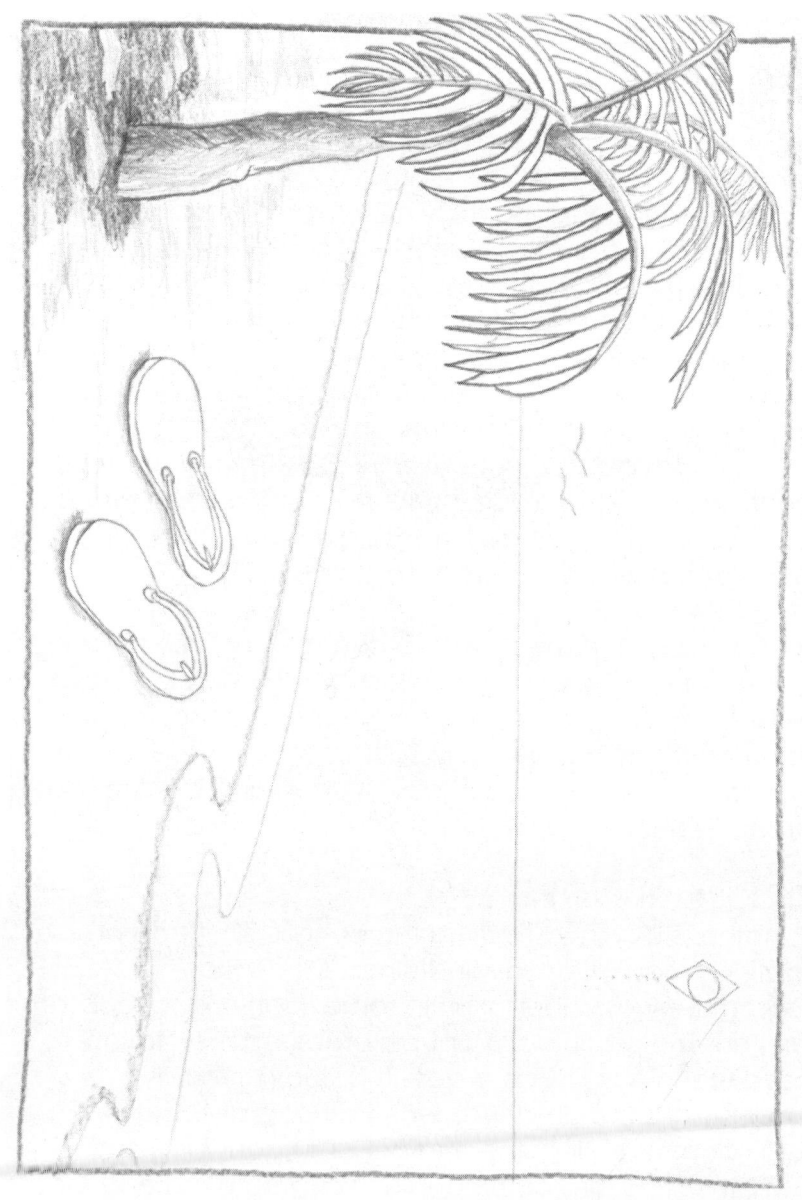

brains very receptive to learning and retaining new information.

Research at Harvard Medical School found significant physical changes in brain structure after only eight weeks of participants undertaking thirty minutes a day of meditation-based practices and techniques. Using MR Images, they were able to confirm that regular meditation increases the grey matter density in different parts of our brains. Significant density changes were recorded in the hippocampus (which is where the hippos camp together), the part of our brain that is vitally important for learning and memory (maybe it's elephants). Increased grey matter density was also recorded in the parts of the brain that are responsible for our self-awareness, our compassionate nature and our ability to be introspective. (7)

We really do have the power to change our brains when we meditate. It's a matter of mind over matter. Mind over grey matter. Bring on the grey matter, for it improves our brainpower and wellbeing. Meditating improves our quality of life and the resonance of our Onesies, and I reckon we can easily slip on the Flip-Flops of Happiness more often.

SCHOOL CURRICULUM
In another study investigating the beneficial effects of practising meditation, a Mindfulness Meditation programme was undertaken in some 'high-risk' schools in America. The participating pupils practised meditation-based techniques twice a day. This study clearly demonstrates the amazing benefits that practicing meditation can bring: the number of suspensions went down, the standard of the pupils' work improved and their attendance at school increased.
If teenagers can meditate, I reckon we can too.

It's fantastic that schools are beginning to embrace meditation as a tool for pupils. A tool to improve their cognitive skills and the standard of their work, a tool to improve self esteem and also a tool to engender more emotional resilience, boosting their whole wellbeing and happiness through the turbulent years of adolescence.

A new study of 7,000 teenagers in the UK, which started in 2016 (3) is going to track the effects of Mindfulness Meditation over a seven-year period and is led by Willem Kuyten, a professor of Clinical Psychology at Oxford University.
This study is to test if mindfulness can improve and boost teenagers' inner resilience to the challenges that often accompany these transformative years.
A cross section of 3,200 children, aged between eleven and fourteen, will receive training in secular mindfulness techniques, such as breathing and walking meditations. These students will be trained every week for 10 weeks. This includes a half-hour lesson at school and twenty minutes daily practice in their own time. A different group of a cross section of 3,200 pupils will continue with the normal Personal Health and Social Education lessons.
Another study, led by Professor Sarah-Jane Blakemore, a neuroscientist at UCL, will be testing 600 eleven-to-sixteen-year-old pupils to investigate if mindfulness training can bring a positive influence in self control and emotional regulation.

So if you, like me, are wishing that meditation lessons had been on offer as well as detentions while we were at school, no worries, because no matter how old our brains are, it's never too late to start meditating. Research (3) has found that meditation not only rebuilds the volume of grey matter in different regions throughout our brains, meditation also helps to preserve this happy brain health through our advancing years. Oh Yes! Bring on the grey matter.

IMMUNE SYSTEM
Practising meditation boosts our immune system and makes our bodies physically more resilient to infection. Because meditation activates our Feel Good Packages, the hormones associated with these feelings are happily hanging around the receptor sites that invading organisms have to use in order to get into our bloodstreams. While our Happy Hormones are busy doing their thing bouncing around our receptor sites, the invaders won't even get a foot in the door.

A study that looked at the correlation between the immune system and meditation found that the participants who took part in an eight-week meditation-based training course had increased activity in the part of the brain responsible for positive emotions. These participants also produced more antibodies to a flu vaccination than the participants who did not meditate. (4)

HEART AND CIRCULATION
Meditation is particularly good for our hearts. The sense of relaxation that spreads throughout our whole bodies when we meditate is accompanied by physical changes to our heart rate and circulation. Once our Relaxation Packages have been activated, our heart rate slows down, the circulation of our blood improves and our hearts and arteries are under significantly less pressure.

One study specifically looked at the levels of lipid peroxide associated (Ref 4) with meditators and non-meditators. High levels of lipid peroxide can be a contributing factor in hardening of the arteries – atherosclerosis – as well as other age-related diseases. They found that meditation lowers the level of lipid peroxide in our blood. They also found that with consistent meditation practice over a period of time, we are able to significantly decrease the thickness in our artery walls. Less pressure on our hearts and less lipid peroxide in our blood means we significantly decrease the risk of heart attacks and stroke.

Impressive stuff, eh? I'm going to hop off now and practise thinning my artery walls for a bit. I've just got to turn down the vmPFC, Maybe Mayhem, turn up the dmPFC, Mostly Manners, switch on Brain Brakes in the Assessment Centre, tune into my Relaxation Package, tell Monkey Mind to go hunt bananas while I slip on the Flip-Flops of Happiness and take my brain on a fifteen minute exotic holiday.

GOOD SLEEP
Because practising meditation is so very good at lowering our blood pressure and our stress levels, it can also help with other conditions that are exacerbated by stress: tension headaches; menstrual tension; ulcers; irritable bowel syndrome; insomnia and other stress- related conditions including visits from the mother-in-law.

Insomnia is the absolute pits. Having enough sleep is quintessential to our good health and happiness. When we're feeling tired, work and normal everyday tasks are more demanding. It's much harder to fully engage in what we're doing, let alone summon up enthusiasm, good manners and creative inspiration. Tiredness often comes hand in hand with irritability, a state of being when we can easily get hold of the wrong end of the stick, take things too personally and end up feeling worse. If we consistontly have trouble sleeping, in the daytime we can end up like the walking dead because brains desperately need their sleep.

Too many unwanted thoughts preventing sleep, or waking in the night, are linked with too much cortisol in our bloodstreams. One of cortisol's jobs is to wake us up in the morning; however, too much of it in our bloodstreams can keep us awake at night and wake us up during the night. Practising meditation reduces muscle tension, significantly lowers cortisol levels and considerably increases the levels of the hormone melatonin. Melatonin is crucial to us all for getting a good night's sleep. Research at Harvard Medical School (4) where long-term insomniac participants were taught meditation-based relaxation techniques, found that after learning these techniques, 75% of the participants could fall asleep within twenty minutes of going to bed. Meditation offers a route to freedom for those who might be stuck in a traffic jam on the road of the walking dead.

SOLACE

Meditation can also offer some respite for those who suffer with depression. Depression is an illness that affects many people in these demanding, stress-filled, sleep-deprived times we're currently living in. If not us, then we probably know someone who suffers in this way, even if we are not aware of it. Depression is a hidden illness that's not readily discussed, even between friends.

Practising meditation can offer some solace from a depressed state of being as it calms the over-talkative, worrying and anxious parts of our minds. Practising meditation soothes our bodies and refreshes our spirit with optimism and resilience. Training the brain to meditate is a powerful tool for taking control and managing the thoughts and symptoms of depression.

A study in America (6) found that practising Mindfulness Meditation was, for some participants, as effective in relieving the symptoms of depression as taking antidepressant drugs.

Meditation cultivates our inner strength and ability to control our thoughts, control the content and delivery of our internal dialogue and control our behaviours. In this way, undertaking a specialised course in Mindfulness Meditation can also be a helpful tool in dealing with addictive behaviour.

WHOLE HEALTH
Practising meditation relaxes the whole body, lowers blood pressure, improves circulation and reduces the risk of us lot experiencing heart attacks and stroke.
Meditating boosts our immune systems and keeps us healthy.
It improves memory, concentration and the ability to learn new information.
Practicing meditation reduces cortisol levels in our bloodstreams, lowering stress, reducing weight gain and improving sleep. No more walking dead.
Meditating switches off the incessant noise of worry and thinking too much, and increases our ability to remain calm in any situation.
Meditation improves relationships, increasing our ability to experience genuine empathy and express our true compassionate nature.
Practising meditation makes our Onesies glow with beauty and radiate happiness into our shared world.
Meditation trains our brains to stay in the only real time that exists.

The relaxed, aware and focused state of being that accompanies meditation is one that can travel with us wherever we go - just slip on the flip-flops and keep walking. With this attitude and awareness firmly woven into the fabric of our Onesies and our footwear we can apply it to whatever we're doing, where ever we are.

The more we practise, the easier it becomes to remain in the present moment. The more we abide in the present moment, the easier it becomes to remain on a happy path of self-actualisation, being the person we really want to be and doing the things we really want to do. The more of us lot who are on this happy path, the more we influence the Collective Happiness Factor.

With a substantial dose of the Collective Happiness Factor in place we can restore the balance desperately needed in our world. We can ensure that the Earth will keep on spinning, keep on orbiting the Sun and keep on trailing in its wake at 483,000 miles per hour,
with us lot still on it.

Meditation is a no-brainer.

Think Now.

CHAPTER NINE

THINK LOVE

If you want happiness for an hour
Take a nap
If you want happiness for a day
Go fishing
If you want happiness for a year
Inherit a fortune
If you want happiness for a lifetime
Help someone else

Old Chinese Proverb

THINK LOVE

If we are wired for anything, we are most definitely wired for love.

It is the Alpha and Omega of life. Love is our origins, our nourishment and our teacher. Many people wonder about the purpose of life, and maybe you, like me, think that the purpose of life is to experience love. Love certainly gives our lives purpose; loving other people, our families, friends, loving ourselves, loving what we do, loving where we live, love brings genuine purpose to our lives and much lasting happiness.

We all want to give and receive love. It's paramount to the human existence and integral to our social nature. Love bonds us together in groups and has been bonding us together since the beginning of humanity. Without love coursing through our blood and bones, we wouldn't have survived this long on our Galactic trail around the universe.

THE WORLD WANTS LOVE

In the world survey that the Happy Planet Index correlated asking countries around the world what they wanted most, after happiness at the top of the list, is love. The whole world wants love, and according to this survey, it's much more important to us than money. If happiness and love are what we all want most, it seems that our governments and the media aren't upholding their responsibilities to represent us fairly and honestly. We want an environment where the expression of these qualities is a priority and active in local communities, in our nations and between nations.

The opposite of love, or the absence of love, is fear. Governments and the media do plenty to perpetuate the belief that we need to be fearful. We are constantly being reminded of reasons to be fearful; the news is mostly comprised of tragic and traumatic events, many of which are generated by war and the adverse effects of extreme weather - plenty of stuff to be scared by.

We are living in a time when the biggest threat to humanity is our changing climate. We know the planet is 100% interdependent and that we all share the same ecology, the same air, the same water and the same roof on our shared greenhouse. In this moment we have the incentive, the skills, the technology and resources to do things differently in the knowledge that our survival depends upon our humility and a willing co-operation. Given all this, it makes the idea of war seem totally absurd. War is absurd at any time, but in our current situation, the perpetuation of war is an even more grotesque misappropriation of life, of money, of energy and of governmental responsibility. Persisting with this way of thinking and the belief that war is necessary is completely out-dated. It isn't what most of us in the world want or believe. Us lot in the world want to pull together, we want to put our Global Brain Brakes into action and redress the balance of the ecosystem that feeds us, the air we breathe and the water we drink. We want peaceful and constructive national and international relations. We want to live happy lives.

SHAMEFUL PERPETUATION OF ARMAMENTS & WAR
In the 1950's, Buckminster Fuller, an American inventor and visionary (1) considered war to be obsolete. One of his thoughts on war was that if the time, money and energy invested in war and the manufacture of armaments was spent instead on living, we would have plenty for everyone on Earth to have a much better standard of living.
It's just as true today. The same equation would work and it's indisputable that we would all be much happier. After

all, the Costa Ricans are doing very well on the Happiness Factor without an army or any weapons.

The amount of money spent around the world on defending ourselves against each other is astronomical. Here's a list of the top ten countries (2) from 'Stockholm International Peace Research Institute' 2015 Fact Sheet (for 2014). The figures, some of which are estimates, are in billions of US dollars:

USA	610.0
China	216.0
Russia	84.5
Saudi Arabia	80.8
France	62.3
UK	60.5
India	50.5
Germany	46.5
Japan	45.8
South Korea	36.7

I'm seriously shocked to discover that the US Government spends more on defence than China, Russia, Saudi Arabia, France, UK, India and Germany **all put together**.

Reading these figures makes me realise that the US Government isn't so much of a 'Super Power' but more of a 'Super Bully-Boy'. It's a bit like the over-sized kid in the playground with the most marbles in his pocket, bullying everyone else to carry on playing marbles even though it went out of fashion decades ago. It seems that the US government is the biggest perpetuator of an out-dated and limiting belief system that upholds fear, war and destruction as a necessary part of life. It isn't. It's an insane waste of money, time, effort and resources directed into destroying lives that could be re-directed into revitalising the planet, saving our own arses and living well together.

Living with a willing attitude of co-operative humility, equanimity and respect.

NUKES ARE BANNED BY LAW
On the 7th July 2017 the United Nations had a big re-think and passed a treaty banning nuclear weapons under international law, with the co-operation of 140 countries. The nine countries with nuclear weapons refused to co-operate, boycotting meetings in order to undermine the legitimacy of the treaty.
The Nasty Nine with Nukes who didn't sign are: USA, China, Russia, France, UK, India, Pakistan, Israel and North Korea. I came across this information via a post on Facebook by Greenpeace International.

MEDIA MANIPULATION
If more of the good stuff were reported on the news, we'd be thinking very differently. I didn't know the UN had banned nuclear weapons until I happened across it on Facebook. Undoubtedly this news was reported in the media, but it wasn't flagged up. On the contrary, the news in the UK is frequently blahing on about the need for us to have nukes without every mentioning that they've been banned under International law.
Most of the news is grim, with a little snippet of something positive at the end. It's not a fair representation as much is happening in the world that is extremely inspiring. Even if 50% of the news material broadcast or reported in the media consisted of positive items, we'd be thinking very differently. If the tables turned completely and the majority of the news was positive, it would radically change our thinking.

If we were being completely bombarded with reports on the successful initiatives already happening to restore our forests, our ecosystem, our rivers, our seas, our roof, together with reports on our local, national and international achievements for happy, sustainable living; if we were constantly shown all this stuff instead we would have completely different individual and collective mind sets.

We would have much happier lives. We would have much healthier families, schools, communities and nations. We would have a much happier world and a much brighter future for our children, our grandchildren and our grandchildren's grandchildren to inherit.
If the news were presented in a more balanced way we wouldn't be caught so easily in a whirlpool of trauma, tragedy and stress from which we escape into devices and vices of distraction because it's all too much.

However, positive news doesn't generate money for the Moguls or the Warlords. News about war perpetuates the belief we need armaments, and armaments generate loads and loads of money.
Positive news changes our way of thinking, happiness re-wires our brains but neither of these life-affirming qualities generates money for the pockets of the Fat Cats in power.

PEOPLE POWER
With more positive news reports, we would be willing to engage more fully. We'd be focusing more on the positive actions and contributions we can make here and now. We'd be more inspired, connected and motivated rather than feeling disheartened and powerless.

We're not powerless. There are billions of us in the world who want more happiness, who want to live healthy, sustainable lives, free of war, conflict, disease and poverty.

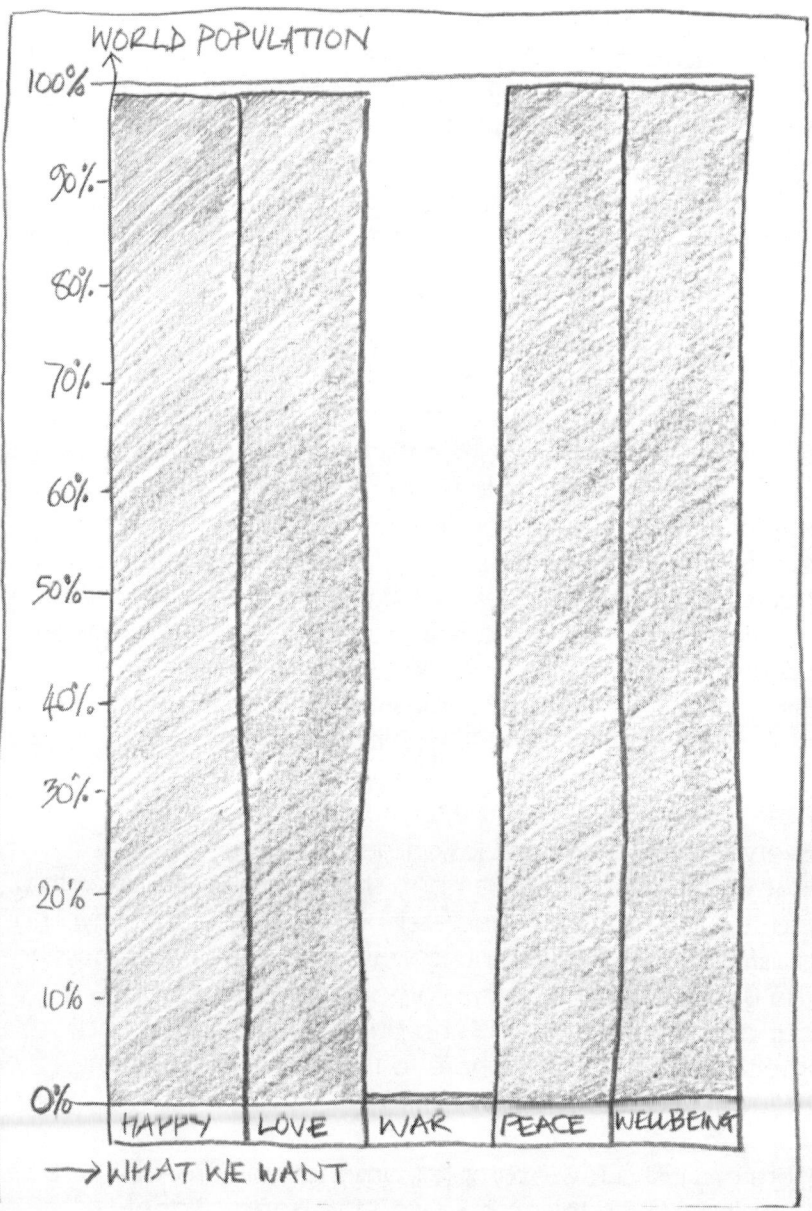

There are billions of us of who value these qualities most in our lives and in the world at large.

Us lot, us billions, we're thinking differently and choosing not to subscribe to the ideas and beliefs that perpetuate war, death, poverty, homelessness, disease and starvation as unavoidable side effects of 21st century living. They aren't. It's a myth. Our world doesn't have to be like this and none of it is unavoidable.

Life right now in the 21st century is exciting precisely because we're thinking very differently. We're co-creating a massive re-think in our collective thinking. It's as if there's a universal thought-programme being collectively and simultaneously downloaded by billions of us, because billions of us are thirsty for change and want to live differently. We're looking in a different direction and want happier, sustainable lives for all. We want peaceful relations between nations, we want equality and fairness, we want community and we all want a planet to live on.

Billions of us are taking People Power into our own hands and downloading the Happy Planet Programme.

HAPPY PLANET PROGRAMME

Everyday there are articles online about ecological innovations and achievements shifting our world back to a place of sustainability and well being for all; from recycling plastic for fuel, kids inventing ways to clean up our mess, to the UN insisting on a shift to organic farming. Change is happening so fast that it's hard to keep up with it and impossible to do justice here to the extent of positive thought in action.

Here are just a few examples, amongst millions, of where People Power is taking hold and the Happy Planet Programme is already installed and running successfully:

Bolivia has made a shining mark in world history by passing a 'Law of Mother Earth' giving rights to the Earth as a living system.
The land, the waterways, and the air are all to be protected from exploitation and contamination. (3)

Researchers in Barcelona have developed a moss-growing, biological concrete that lowers CO_2, insulates and regulates heat and also creates an eye-catching vertical garden. (4)

A city in Alberta, Canada, is the first in its country to completely eliminate homelessness from amongst their citizens. (5)

A city park filled with hundreds of edible plants, fruit trees, vegetable plants and herbs, free to all, is being planned by Seattle council and it will be the first 'food forest' in the US. (6)

The Independent reported in October 2015 on a study by 'Bloomberg New Energy Finance', that Wind Power has gathered sufficient momentum to cross an economic threshold, and is now the cheapest way to produce electricity in the UK and Germany.

Manoj Bhargava, an Indian entrepreneur living America, hard working and self-made billionaire, is really on the case. He has established the 'Knowledge Medical Charitable Trust' and 'Rural India Trust' supporting many initiatives including building hospitals for the poor and education for disadvantaged women in rural areas. He has also co-founded 'The Hans Foundation' which supports hundreds of charitable organisations across India. In 2012 he joined the

Giving Pledge and pledged to re-direct 99% of his wealth to fight against poverty and find solutions for some of the world's fundamental problems. His 'Invention Shop' is dedicated to inventing and manufacturing sustainable, low cost or free technology to provide water, electricity and improved healthcare for the poor. Manoj and his team are preparing 10,000 'Free Electric' bicycles that generate electricity, ready for people in India by March 2016. The recently released short film about his projects already underway, is called 'Billions in Change' (October 2015). Within 12 days of its release, 19,000 people, plus one more, pledged to volunteer. (7)

Wikipedia informs me that the Giving Pledge was kick-started in 2010 by Bill Gates and Warren Buffet, to encourage other billionaires to pledge some of their wealth to philanthropic causes. As of August 2015, 137 billionaire or former billionaire (on account of giving their money away) individuals or couples have signed the pledge. Currently, the significant majority of these philanthropists are American citizens, and the number of billionaires joining this pledge is growing swiftly all the time.

There really are billions of us, and billionaires, thinking very differently.

POSITIVE CHOICE
Where possible we are choosing products, manufacturers and companies that have sustainable integrity at the heart of their beliefs and ethos. Making these positive choices in our thoughts and beliefs, as well as in our spending habits, has a profound effect on each other and the world at large. Even though we are susceptible to the manipulative powers of advertising, together we have a huge influential power over the commercial markets. If we no longer buy

into these ideas or no longer buy the products, the old beliefs will wither away.

Nations influence other nations in taking positive steps for the common good. For example, billions of us don't want to eat genetically modified food and many countries around the world have banned, or have strict restrictions in place on, Monsanto's GMO's and their 'Roundup' pesticide. We don't want genetically modified food and we don't want to eat glyphosate either, a chemical in Roundup that has made its way into much of our food and into our bloodstreams. It's a chemical that's linked to cancer and other health issues in humans as well as killing the bees, other pollinators and other species of wildlife in devastating proportions. Together we're demanding something different. We're demanding normal, uncontaminated food, a demand that is fundamental for the survival of us all.

The power for change really is in our hands, our hearts and our minds.

We are living in a pivotal point in human evolution and despite how things are portrayed in the media we've already chosen to swing the pendulum back towards common sense. We're already thinking very differently and we're already creating a global community of guardianship.

ROLLING WAVES

The momentum for change is growing fast and is directly, and indirectly, fuelled with rolling waves of fresh inspiration from us lot and the quality of our thoughts, our words and our Onesies.

Although we can't change anyone else, as much as we might like to sometimes (a few politicians spring to mind), we can change ourselves, and in so doing, we positively influence each other with enthusiasm and love for life.

We can begin to change the way we think about ourselves, which in turn will influence the way we think about, and interact with, others. Remind ourselves it's up to us; our responses bring either happiness or unhappiness. We can Think Thanks on waking each morning. We can write different shopping lists and if the sugar virus is trailing behind, we can anchor ourselves in the knowledge that it's a toxic chemical leading straight to the cemetery. We can Think Smile, especially in the face of adversity. We can Think Creative, particularly with the mundane. We can Think Now and breathe fresh life into each moment. We can allow ourselves to Think Love.

As our Onesies start to send a different message out to our families, partners, friends and colleagues at work, these subtle differences are perceived and others respond accordingly.

We might not be able to change other people directly, but we can influence each other for the better, for happier molecules all round.

We have an effect on each other all the time with our varying moods and feelings, our thoughts, our words and our actions. Love is the most powerful force we have to facilitate our personal change and cultivate more inner peace, contentment and happiness. It's also the most powerful force we have to influence and encourage each other.

Embracing our own weaknesses, contradictions and inconsistencies and accepting ourselves exactly as we are, tempers the voice of King Critical, giving us a different perspective on those parts of ourselves. From a compassionate viewpoint it's much easier to think differently, easier to let go of the thoughts, feelings and beliefs that impede happiness and easier to implement the changes we want for ourselves. The more we cultivate an attitude of kindness towards ourselves and accept 'the good, the bad, and the ugly', the easier it becomes to hold onto an attitude of kindness towards others. After all, everyone's got foibles and inconsistencies.

CAPTAIN HEART-LOVE

As it turns out, Brain isn't actually the boss. It transpires that the brain acts more like a Lieutenant, as it's the heart that operates as Captain and is in charge of navigating our way through the voyage of life.

I admit it's not really news, we know intuitively how important it is to acknowledge and follow the truth that nestles in the dreaming of our hearts. Science now confirms this (8) and what sages of old have been saying for a time period of at least nine and a half donkeys: that wisdom, truth, the peace of right action, of kindness, acts of beauty, acts of compassion, joy, appreciation, creativity, happiness and love all spring forth from the energy and the power that is in our hearts.

The energy that our hearts generate is quite astonishing.

The heart's electrical field amplitude is 60 times greater than that of the brain and it permeates every cell in the body.

The heart's electromagnetic field is 5,000 times stronger than the electromagnetic field generated by the brain.

We might not be able to see these electrical and electromagnetic fields with the naked eye, but we can't see the blind spot that's right in front of our eyes either (our brains fill the gap in for us), and both are nonetheless very real. These fields, in conjunction with our thoughts, feelings, values and beliefs, create the Onesy that surrounds each of

us, operating as antennae through which we interpret the world.

THE HEART-BRAIN

Leading the way in this new field of 'neurocardiology', are the research team at the 'Institute of HeartMath', a non-profit, research and educational institute committed to helping people develop the skills to reduce stress, manage emotions, and cultivate resistance and fresh energy for happy and healthy lives.

Much of the research by Rollin McCraty PhD, Raymond Bradley PhD (9) Dana Tomasino BA and others at HeartMath has focused on the heart and brain communication. Their studies highlight how this constant dialogue affects our awareness, our consciousness and the way in which we think, perceive and respond to the world around us. It demonstrates that our hearts do a great deal more than the magnificent job of pumping blood around our bodies about 100,000 times a day.

Their research shows that the heart has its own 'brain'. It's a brain with 40,000 sensory neurons continuously transmitting information to the cranial brain. Surprisingly, on a daily basis the heart-brain sends more information to the brain in our heads than the other way around. Our heart really is Captain in Command.

Heart-brains talk to cranial brains in four different ways: via the nervous system, via hormonal information from the hormones produced in our hearts, via biochemical information through our varying blood pressure and the gaps between each heart-beat, and via energetic information picked up by our antennae within the strong electrical and electromagnetic fields of our Onesies.

The heart's electrical and electromagnetic fields, together with the 40,000 neurons in the heart-brain, form our initial response to absolutely everything and everyone we encounter and experience. Our hearts respond first. The heart then sends this information, encoded by our response, to the brain in our heads. The cranial brain in turn responds to Captain's orders with various forms of brain activity. The nervous system in our hearts, with its 40K of brainpower, provides a means for the heart to learn information, remember stuff, and make executive decisions without consulting the Lieutenant. They've done shedloads of experiments at HeartMath and have found that the messages Captains are constantly supplying to Lieutenants influence the bits of the brain involved with perception, cognition and processing emotions.

They've also discovered that our brain's rhythms are designed to fall in line with the rhythms of our beating hearts.

There's a significant difference in the beating patterns and intervals that our hearts generate when we experience different emotions. As we know, anger can make our hearts go bonkers and beat extremely fast as our blood pressure rises in accordance with our temper. Similarly, other negative emotions cause irregular, unstable and disjointed patterns in the rhythms of our hearts and Captain's orders. Positive emotions set our hearts beating in a very different way. They beat out stable, regular and coherent patterns with messages that permeate through to our electric and electromagnetic fields, which in turn respond accordingly.

These clever peeps at HeartMath have actually been able to measure changes in the heart-fields using a technique called 'spectral analysis'. I don't think they offer a ghost-busting service though. They've been able to confirm that when we're experiencing a period of sustained positive emotion, our hearts, brains and bodies operate in a happy, coherent way with better efficiency and congruity. This state of being is connected with a significant reduction in the Mostly Mayhem aspect of our internal dialogue. It lowers stress, restores a feeling of emotional balance, improves the Lieutenant's ability to think clearly and do its job of keeping brain and body ship-shape. This state of congruity improves intuitive perception and sensitivity as our Mostly Manners package is activated when in a coherent and harmonious state of being.

When our hearts, minds and bodies are in a state of harmonious coherence, our Onesies positively glow and bleep with happiness, impacting on the heart-fields belonging to anyone else we encounter.

There's no escaping the Onesy Fields. We can't switch them off, but we can certainly fine-tune them to the frequency of happiness.

The findings at HeartMath confirm that much of the information we perceive and understand about each other is in the form of energetic information. We're responding to each other on an energetic level all the time and this is where our real power lies to influence and encourage each other for the better. Think Happy; radiate happy. Think Love; radiate love.

KINDNESS

Just as smiling at a friend elicits a smile in return, good vibes generate more good vibes in those around us. We can train the brains in our hearts and in our heads to default to the Good Vibe preference and upgrade our Think Love Programmes to start running the best possible version of ourselves. We could listen to our heart's desires more often. We can delete limiting belief programmes and download 'The Beginner's Guide to Loving Kindness' and learn how to cultivate a gentle attitude towards ourselves and each other, instead of battling against our own and other peoples' insecurities and inconsistencies. As we begin to cultivate a Think Love attitude, the nature of our emotions and our power to think differently becomes more apparent. Much of what we experience in life is transient and changing, emotions passing through; love is consistent and a state of being that provides real nourishment and the ultimate comfort for mind, body, Captain and soul.

Kindness is one way Love loves to express itself. Kindness was encoded into our ancestral genome as a way of assuring the survival of the whole group, as the greater the emotional connections between people in the group, the bigger the chances of survival. Kindness has a sticky quality that works as a social fixative; we all appreciate kindness and remember those who are kind to us. Kindness plays a very important role in our human existence and it's a quality that can be active and upgraded at any time.
Like laughter, kindness is infectious. It's good for the Captain in Command, slows ageing, makes for better relationships and keeps us feeling good and happy.

ROLLING KINDNESS EFFECT
An act of kindness, random or otherwise, really does inspire others to follow suit. David R Hamilton PhD, American author of 'Why Kindness is Good For You' (10) says that the ripple effect of kindness spreads to three degrees of separation, to our friends' friends' friends.
A scientific study followed the trail of an amazing act of kindness after an anonymous person, twenty-eight years old, decided to play the Ace of Hearts and donate one of their kidneys to a clinic. It set a Rolling Kindness Effect in motion where relatives of those who'd received a kidney donation, followed suit and made the same life-saving gift to those in need. In this case, the Rolling Kindness Effect rolled all the way across America, north, south, east and

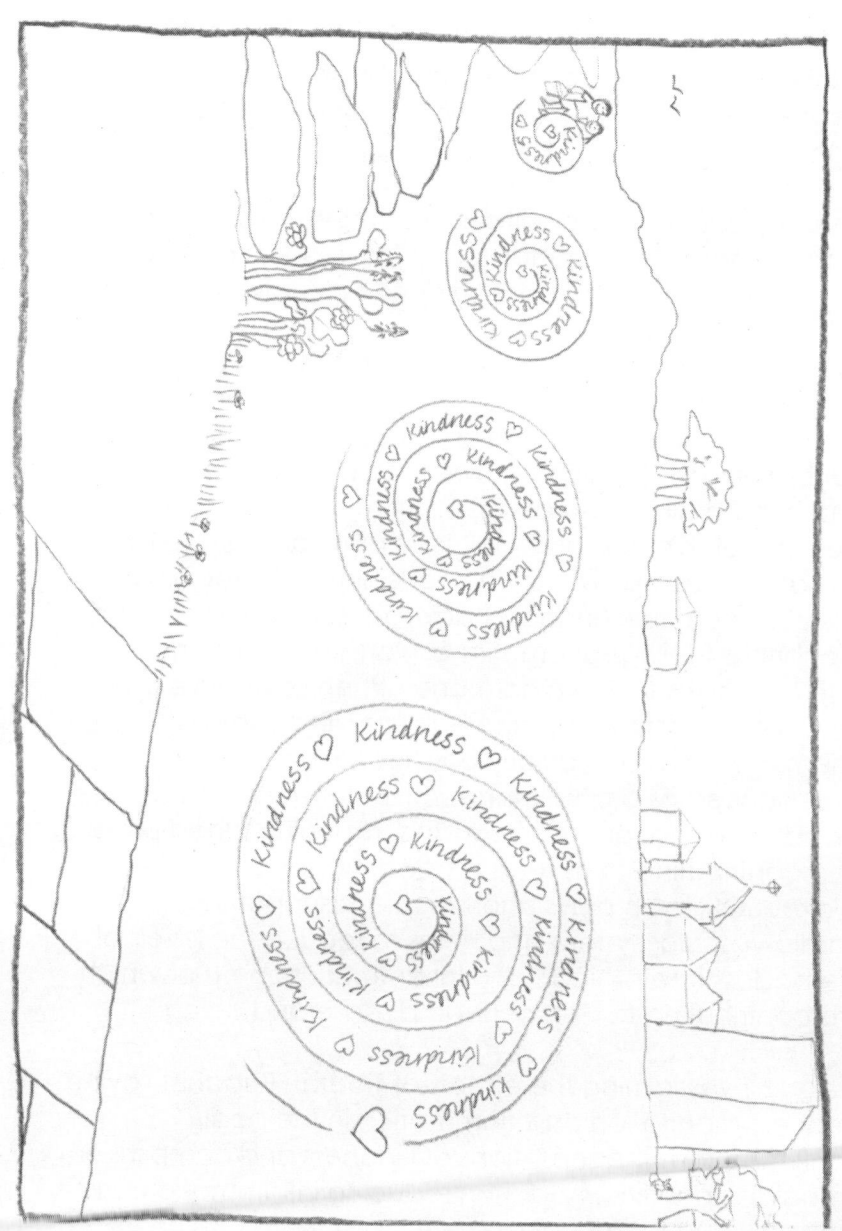

229

west. From that first act of ultimate kindness, the lives of ten people were totally transformed with the gift of a new kidney.

Although we might not witness the furthest reaches of the Rolling Kindness Effect that an act of kindness can generate, it's heartening to know that kindness spreads more kindness in this way.
Kindness comes in tandem with good feelings of emotional warmth as the Kindness Package releases different hormones into our bloodstreams. The level of dopamine in our brains increases. Dopamine is an endogenous opioid and is the brain's natural version of morphine or heroin. The increase of this hormone makes us feel good and this sensation is also known as 'helper's high'. Oxytocin is also released from the Kindness Package, a bonding hormone that brings feelings of emotional warmth and connection with others. It's the love hormone. When oxytocin is present in our blood, it triggers the release of nitric oxide, a chemical that is particularly good for our hearts. Nitric oxide is recognised as a cardio-protective hormone as it causes blood vessels to relax and expand, reducing blood pressure and contributing to us feeling good, happy and more relaxed. Oxytocin plays another role in supporting the cardio-vascular system and helps to reduce the levels of the badass free radicals and inflammation, both of which are contributing factors of heart disease and aging.

A study investigating the effects of Tibetan Buddhist Loving Kindness meditation on inflammation in the body, confirmed that the sensations of kindness and compassion do have a positive impact on reducing inflammation. This is probably due to the calming effect on the vagus nerve. Amongst other jobs, the vagus nerve regulates the heart rate, regulates inflammation levels and there is some evidence to suggest that there is a strong link between the activity of this nerve and the part of our brain that accesses a state of compassion.

Kindness sticks us together, makes us feel good and makes us feel happier. An act of kindness can completely transform someone else's day, and in such a way as to set the Rolling Kindness Effect into action. Kindness is the Queen of Hearts, is extremely good medicine for the Captain in Command and helps to keep his pecker up at all times. That's the British English version of pecker, not North American.
(Or possibly both.)

GENEROSITY
Another way Love loves to express itself is through acts of generosity, volunteering our time and energy to help others and by donating to charities. These acts of generosity bring significant benefits to those we are helping and brings us many positive feelings of happiness.
Scientific studies (11) examining participants' brains when engaged in an act of generosity have shown that there is a direct link to the pleasure centre regions of the brain. This part of our brain is particularly happy when we make a donation from our own purse.
In one study, which gave away money to random people, found that those who were asked to donate some of it to charity reported higher levels of happiness than those who were told to keep the money for themselves.
Donating money to charity and volunteering our time and energy go a long way to strengthening our sense of connection to others in our communities and the world at large. Research by the Charities Aid Foundation has found those in the top 9% of us in the UK who volunteer and

donate to charity, are significantly more likely to engage with our neighbours and say hello than those of us who don't donate or volunteer at all. We could all say hello to our neighbours, couldn't we?

Adam Pickering has written many inspiring and informative articles on the 'Future World Giving' website, which provided the basis for some of the above as well as these next bits of facts and figures.

The World Giving Index (WGI) analyses charitable behaviour in 135 countries and calculates a score based on an average of three types of giving behaviour in a typical month: donating money to charity, volunteering time, and helping a stranger. This analysis shows there's a strong link to increased wellbeing in the countries where many people regularly donate to charity, volunteer and help strangers.

Countries in the top ten of the 'Gallup-Healthways Global Well-Being Index', have an average score on the WGI that is 7% higher than the global average WGI score.

Similarly, on the Happy Planet Index, the average score between the highest-ranking nations is 11% higher than the global average of World Giving. These acts of generosity aren't necessarily linked to having a fat wedge in the bank. Myanmar, formerly known as Burma, which has a ranking of 150th on the 'Human Development Index' (which calculates a score based on life expectancy, education and per capita income), took 1st place in the 2014 World Giving Index, along with the USA. Furthermore, only five countries in the G20 (the Group of 20 Major Economies) are to be found in the top 20 on the WGI. Eleven G20 nations are ranked lower than the top 50 on the WGI, and surprisingly, three of these countries are lower than the 100 mark.

The fluctuations in global giving trends are linked with the economy and in particular, linked with our innate human desire to help each other through crisis, conflict and tragedy. We're wired to help others in need, whether in our communities, in neighbouring countries or in countries far away. The advent of the world-wide-web has brought everywhere much closer and witnessing others suffering conflict or disaster from the comfort of our homes has an immediate and direct impact on us.

The Indian Ocean Tsunami in 2004 caused an unprecedented reaction around the world and the response has been described as 'the most generous and immediately funded humanitarian response in history'. (12) The tsunami had a particularly profound impact on us lot in the UK. We were the first country to respond with financial aid and together, the UK public immediately donated an astonishing amount of money, raising 10 million pounds in just 24 hours. This act of generosity broke the Guinness World Record for the most amount of money donated online in 24 hours, and is still currently the record appeal. The UK Disaster Emergency Committee received a total of £392 million in donations from us lot, the UK public. This is the highest contribution to a single appeal since 1963 when the committee of Non-Governmental Organisations first formed. The UK also made a further donation of £89 million that came from the money we put in our government's purse. Collective acts of generosity like this make me feel proud to be part of such a generous and caring nation.

VOLUNTEERING
The UK sits in 8th place out of 135 countries in the 2014 World Giving Index and part of this equation is how we also volunteer our time and energy. The New Economic Foundation have identified a set of evidence-based

actions which increase our level of wellbeing, and volunteering has been proven to be an important part in keeping our happiness levels intact.

Volunteering our time and energy brings many benefits to those we are helping and brings us many rewards through connecting with others, strengthening a sense of belonging and blasting our bloodstreams with healthy, happy hormones.

Belonging is part of our ancestral mix and a part of our nature. We all want to belong and having a sense of belonging is just as vital today as when it was encoded into our genes.

Our happiest times are probably those we have spent with others, when our sense of belonging is 100% intact. With a Belonging Programme up and running through our lives, happiness, good health, expressions of love and the bonding hormone Oxytocin all flow along more easily. It's when we feel out on a limb, when we feel isolated and disconnected, that life becomes harsh, and happiness can leak from the pockets of our Onesies just as easily as sand will slip through our fingers. Above all else, we are a social species and need connection with others, so if happiness is leaking from your pockets in this way join something, meet new people and get a lover.

There are many ways we can volunteer and donate time and energy. In doing so, we have greater levels of connection and personal happiness and the opportunity to express and experience one of the most fundamental ways of being human: helping each other.

Our families, partners, friends, neighbours and communities form the bedrock of our lives. Taking time to keep these connections in place, growing and flourishing brings a deep enrichment to our lives and ensures that our Captains, and our Onesies, are stuffed with love and happiness.

WE ARE THE CHANGE
We are wired to keep re-thinking, wired to keep learning and wired to adapt to change. With both hands firmly on the steering wheel, we can train our brains to default to Think Happy and collect words and options that enable us to think differently.

We can train our brains not to go Kangaroo Jumping to wrong conclusions and assumptions. We can summon the courage to open the door to the Cupboard of Doom and clear out limiting beliefs, ditch the thoughts and Baggage of Stuff that doesn't work; that no longer serve our hearts or the heart of humanity. With our flexi-wiring, we can re-write the text in our Personalised Guide Books with words of strength and expressions of love.

We can re-think the sugar thing that's more addictive than cocaine and perniciously destroying millions of lives. We can eat our way to happiness from a cornucopia of goodness, train our shopping lists to behave at the supermarket and train our brains not to get caught in a sticky web of persuasive and manipulative marketing that serves only to line its own purse with the wool it's fleeced off our backs.

We can Think Thanks much more often. We can buy less stuff and eat the leftovers in the fridge while reminding ourselves that if everyone in the world consumed resources at the same rate as us in the UK, we would need three planets.

With the Think Thanks programme running in the neural networks of our hearts and minds, we can restore our collective attitude to gratitude. If we allow the peace of gratitude to keep flowing through our bloodstreams, it will radiate out of our Onesies onto our families, partners, friends, neighbours, colleagues and communities.

We can Think Smile more often, activate the Happy Halo effect in our brains and leave a gentle trail of happiness lingering in air. After all, smiling is ridiculously easy; all we have to do is move our face about a bit.

Laughter fires all five cylinders at once and ensures we get a flurry of Happy Chemical texts landing in the inbox. Laughter is 'manna from heaven' and fills our minds, bodies, Captains and souls with the best medicine on Earth.

We can Think Creative, switch off devices and vices of distraction more often and make time to download the latest versions of Purpose, Achievement & Competence, Inspiration & Motivation, Resilience & Happiness XX. Anything goes; it is the act of being creative that fills us with these qualities. Ideas generate more ideas and inspiration is infectious. We could sign a contract with ourselves or sign up for an evening class; there are plenty out there to choose from.

We can train our brains to Think Now. It's the only real time there is and it's where happiness lives.

We can slip on the Flip-Flops of Relaxation and take our brains on holiday every day. A power-packed fifteen-minute flight to everywhere and nowhere simultaneously. On-board services include: relaxation, improved wellbeing, stress reduction, worry reduction, improved sleep, improved brainpower, automatic RAM upgrade, awareness upgrade, restored compassion-ability-drive, and happiness filters shining like new. All this, in just fifteen minutes, and completely free of charge!

We can follow Captain's orders and Think Love. Captains are at their absolute best when we Think Love. We could kick-start the Rolling Kindness Effect into action any day we like, every day if we want to. We could say hello to our neighbours, volunteer in our communities and nurture friendships and relationships that define us as the social, collective species that we are. We can cultivate humility, equanimity and respect, and go home with happiness stuffed into the pockets of our Onesies. Think Happy and our Onesies will light up this world with rays of trust, inspiration and happiness.

Aristotle said 'Happiness depends upon ourselves', which is just as true today, two thousand, four hundred years down the line.
What has changed since Aristotle's day, however, is the perilous state of our fragile ecology and the urgency for change.
Right now, happiness depends upon all of us.
We have a responsibility to think differently about ourselves and how we engage with each other, and a responsibility to restore and replenish the Earth we walk upon. A crucial part of upholding these responsibilities is to be happy.
Happiness is what the world wants most and it's what the world needs most.
Happiness is a tree of life that bears the vital fruits of inspiration, innovation, connection and collective action.
As individual as our bodies might seem, our Onesies are constantly in touch with each other and constantly in touch with the rather large XXXXX sized Onesy that connects life together. There is no escape route. We are all connected to the Onesy Web that weaves all life together. It's a web that we can directly influence for the better, and right now there's an urgent need for us all to use our influence and Think Happy.

We know what it's like to be individual people, individual nations; we've examined our differences for long enough.

= 1 BILLION HAPPY PEEPS

AS YOU CAN SEE, THIS DIAGRAM CLEARLY REPRESENTS 99.99% OF THE WORLD POPULATION.

THIS DIAGRAM ALSO ILLUSTRATES MORE SOULS THAN OUR CURRENT POPULATION, WHICH IRREFUTABLY DEMONSTRATES THAT WE HAVE UNSEEN HELPERS NUDGING US INTO RIGHT ACTION AND WORD HAS IT THAT WE ARE GONNA SAVE OUR ASSES AFTERALL.

Focusing on what unites us as one human race puts us in the frame of mind to realise our common ground; we are just one species of mammal among many thousands of mammals, on one rather beautiful and ingenious planet. The Earth is not ours to keep, we are merely born of it, made from its elements and our passing through is but a heartbeat in eternity.

International differences bring us an astonishing wealth of skills, aptitudes, abilities, vision and inspiration. They bring us the stunning beauty of different coloured eyes, hair and skin. And importantly, they bring us different dance moves. Differences in faith provide many roads, leading many hearts to abide in the same territory of love.
Many religions of the world are re-thinking and striving to unite in order to pave a different way forward, pave a path of world peace.
Governments are slowly responding to what the citizens are saying, or they are slowly crumbling in the force of our innate desire for common good and common sense.

And most importantly, there's us lot. We have a collective responsibility to get a wiggle on, stick a smile on our faces and Think Happy.

Think Happy Planet.

THINK HAPPY PLANET – SOURCES

CHAPTER 1 – THINK HAPPY

Ref 1: PM's Speech on Wellbeing – 25 NOV 2010 – www.gov.co.uk

Ref 2: Office of National Statistics: Measuring What Matters – www.ons.gov.uk
The What Matters to You? Survey – nationalwell-being@ons.gov.uk

Ref 3: Nic Marks – Talk on TEDX – www.happyplanetindex.org

Ref 4: Wikiprogress Statistics – www.stats.wikiprogress.org

Ref 5: Laura Stoll – 21 JUNE 2012 – www.neweconomics.org

CHAPTER 3 – THINK THANKS

Ref 1: In Praise of Gratitude – 1 NOV 2011- Harvard Health Publications www.harvard.edu

Ref 2: Huffington Post – 29 MAY 2015 – www.huffingtonpost.co.uk

Ref 3: WRAP – Waste Resources Action Programme – www.wrap.org.uk

Ref 4: France Supermarket Food Waste Law Piles Pressure on The UK – Ian Quinn – 22 MAY 2015 – www.thegrocer.co.uk

Ref 5: UK Tops Chart of Food Waste – Ami Sedghi – 22MAY 2015 – www.theguardian.com

Ref 6: Tesco Teams Up With Fairshare Charity to Reduce Food Waste – Sarah Butler – 4 JUNE 2015 – www.theguardian.com/business

Ref 7: The Truth About Obesity: 10 Shocking Things You Need to Know – Sarah Bosely – 23 JUNE 2014 – www.theguardian.com

Ref 8: Chris Smythe, Health Editor – 30 MAY 2015 – www.thetimes.co.uk

Ref 9: The Andrew Marr Show – 31 MAY 2015 – www.bbc/news/health

Ref 10: Public Health Responsibility Deal – Department of Health – www.dh.gov.uk

Ref 11: Taking a Bite Out of Climate Change – Anna Lappe (accent on the e) Adapted from her book " Diet For A Hot Planet' – www.sustainabletable.org

CHAPTER 4 – THINK SUGAR? THINK CHEMICAL

Ref 1: Sweet Poison: Why Sugar is Ruining Our Health – Victoria Lambert – 11 DEC 2014 – www.telegraph.co.uk

Ref 2: Sugar Love (A Not So Sweet Story) – Rich Cohen – AUG 2013 – www.nationalgeographic.com
Their reference: Sidney Mintz – "Sweetness and Power: The Place of Sugar in Modern History' published by Penquin Books Ltd – 1985

Ref 3: Britain is Built on Sugar: Our National Sweet Tooth Defines Us –
13 OCT 2007 – www.theguardian.co.uk - no author shown

Ref 4: 10 Disturbing Reasons Why Sugar is Bad For You – Kris Gunnars – www.authoritynutrition.com - no date shown

Ref 5: Just HALF a Can of Coke is MORE Than the New Daily Sugar Guidelines Backed by Scientists who Recommend Just Three Cubes a Day - Lizzie Parry – www.thedailymail.co.uk

Ref 6: How Flavoured Water Contains More Sugar Than Cola – Susan Quinn – 31MAR 2015 – www.telegraph.co.uk

Ref 7: The Harmful Effects of Sugar on Mind & Body – www.macrobiotics.co.uk/sugar.htm - no date or author shown

Ref 8: How Does Sugar Affect Your Brain? Turns Out in a Very Similar Way to Drugs & Alcohol – Dana Dovey – 25 JULY 2014 – www.medicaldaily.com with reference to Dr Nicole Avena

Ref 9: This is What Sugar Does to Your Brain – Carolyn Gregoire –
6 MAY 2015 – www.huffingtonpost.co.uk

Ref 10: UCLA Study Shows High Fructose Diet Sabotages Learning, Memory – Elaine Schmidt – 15 MAY 2012 – www.ucla.edu/releases/this-is-your-brain-on-sugar-ucla-233992
First published in Journal of Physiology

Ref 11: www.actiononsugar.org

Ref 12: www.bbc.co.uk/news/health

CHAPTER 5 – THINK HAPPY FOOD

Ref 1: Brain Food: Nutrition Tips for a Healthy Brain –
www.brainandspine.org.uk

Ref 2: Brain Food: 6 Snacks Good For The Mind – Olivia
Goldhill –
23 JAN 2015 – www.telegraph.co.uk

Ref 3: The Best Foods For Your Brain – www.prevention.com

Ref 4: You Are What You Eat – Kirsten Hartvig ND & Dr Nic
Rowley
Published: 1996 BCA by arrangement with Judy Piatkus Ltd.

Ref 5: 10 Proven Health Benefits of Coconut Oil – Kris
Gunnars –
JULY 2013 – www.authoritynutrition.com

Ref 6: 10 Foods To Boost Your Brain Power –
www.bbcgoodfood.com

Ref 7: 10 Proven Health Benefits of Turmeric and Curcumin –
Kris Gunnars MAR 2014 – www.authoritynutrition.com

Ref 8: Green Tea Isn't Just Good For Your Heart It's Good
For Your Brain Too – Claire Bates – 6 SEPT 2012 –
www.thedailymail.co.uk

Ref 9: Need Protein? Here Are 9 Amino Acids Found
Abundantly In Plants – Health Monster - Heather McClees –
www.onegreenplanet.org

CHAPTER 6 – THINK SMILE

Ref 1: Laughter is The Best Medicine – The Health Benefits of Humour and Laughter – www.helpguide.org - no author shown

Ref 2: The Untapped Power of Smiling – Ron Gutman – guest at www.forbes.com

Ref 3: Laughter: The Glue of Humanity? – Kirsten Coueleskie – www.serendip.brynmawr.edu

Ref 4: Negative Effects of Cortisol – Dr Brent Barlow – 12 JULY 2011 – Natural Health News – www.castanet.net

Ref 5: Norman Cousins - www.wikipedia.org

Ref 6: One Smile Can Make You Feel A Million Dollars – MARCH 2005 – www.thescotsman.com - no author shown

Ref 7: How Smiling Changes Your Brain – Vivian Giang – www.fastcompany.com

Ref 8: The Science of Smiling – A Guide to The World's Most Powerful Gesture – Leo Widrich – APRIL 2013 – www.bufferapp.com

Ref 9: How Laughter Works – Marshall Brain – www.howstuffworks.com

Ref 10: Emotion: The Key to the Mind's Influence on Health – Paul Mc Ghee PhD – www.laughterremedy.com

Ref 11: Happily Ever Laughter – Peter Doskoch – JULY 1996 – www.psychology.com

CHAPTER 7 – THINK CREATIVE

Ref 1: Astronomical Society – www.astrosociety.com

Ref 2: Not So Tortured Artists: Creativity Breeds Happiness – Tom Jacobs FEB 2014 - www.psmag.com

Ref 3: How Our Brains Work When We are Creative: The Science of Great Ideas – Belle Beth Cooper – NOV 2013 – www.bufferapp.com

Ref 4: The Real Neuroscience of Creativity – Scott Barry Kaufman
AUG 2013 www.scientificamerican.com
Also: Brain Scans of Rappers Sheds Light On Creativity – Daniel Cressey
NOV 2012 – www.nature.com

Ref 5: The Emotional Life – Creativity – www.pbs.org - no author shown

Ref 6: A Creative Life is A Healthy Life – Amanda Enayati – MAY 2012
www.ccn.com

Ref 7: www.farnhamstreet.blog - Shane Parrish SEPT 2013 with reference to: A Technique For Producing Ideas by James Webb Young – first published in1939
CHAPTER 8 – THINK NOW

Ref 1: www.dailymail.co.uk - article by Ellie Zolfagharifard – 28 JAN 2014

Ref 2: Astronomical Society – www.astrosociety.org

Ref 3: Mindfulness Study to Track Effect of Meditation on 7,000 Teenagers – Robert Booth – 15 JULY 2015 – www.theguardian.com

Ref 4: Physical and Mental Benefits of Meditation – www.meditationsUK.com - no author shown

Ref 5: This is Your Brain on Meditation – Rebecca Gladding MD –
22 MAY 2013 – www.psychologytoday.com

Ref 6: 7 Ways Meditation Can Actually Change The Brain – Alice G Watson – FEB 2015 – www.forbes.com/business

Ref 7: Harvard Study Unveils What Meditation Literally Does to The Brain – Arjun Walia – 11 DEC 2014 – www.collective-evolution.com

CHAPTER 9 – THINK LOVE

Ref 1: The Buckminster Fuller Institute – www.bfi.org

Ref 2: List of Countries by Military Expenditure – Stockholm International Peace Research – www.wikipedia.org

Ref 3: Bolivia passes "Law of Mother Earth" – 14 MAY 2014 – www.theearthchild.co.za

Ref 4: Moss Growing Concrete absorbs CO2, Insulates and is also a Vertical Garden – Swikar Oli – 24 OCT 2015 www.theplaidzebra.com

Ref 5: The First Canadian City to Eliminate Homelessness – Terry Turner – 7 OCT 2015 – www.goodnewsnetwork.org

Ref 6: Seattle Planning City Park Filled With Edible Plants
www.thehungersite.com no author shown

Ref 7: Manoj Bhargava – Billions in Change
www.billionsinchange.com
News report from India TV USA (www.indtvusa.com)
On www.youtube.com

Ref 8: 7 Scientific Reasons Why You Should Listen to Your
Heart (Not Your Brain) – Dr Joel Kahn – 16 DEC 2013 –
www.mindbodygreen.com

Ref 9: The Heart Has Its Own Brain & Consciousness by:
Rollin McCraty PhD, Raymond Trevor Bradley PhD & Dana
Tomasino BA www.heartmath.org
Article - www.in5D.com

Ref 10: 5 Beneficial Side Effects of Kindness – David R
Hamilton PhD
JUNE 2011 (Author of 'Why Kindness is Good For You')
www.huffingtonpost.com

Ref 11: Giving Makes Us Happy: Charity Improves Wellbeing
for Donors Too – 20 MARCH 2015 - Adam Pickering -
www.futureworldgiving.org

Ref 12: Indian Ocean Tsunami: 10 Years On
www.theguardian.com